Journey Across the Life Span:
Human Development and Health Promotion

Birth is a beginning
And death a destination.
And life is a journey:
From childhood to maturity
And youth to age;
From innocence to awareness
And ignorance to knowing;
From foolishness to discretion
    And then, perhaps, to wisdom;
From weakness to strength
Or strength to weakness—
    And often back again;
From health to sickness
    And back, we pray, to health again;
From offense to forgiveness,
From loneliness to love,
From joy to gratitude,
From pain to compassion,
And grief to understanding—
    From fear to faith;
From defeat to defeat to defeat—
Until, looking backward or ahead,
We see that victory lies
Not at some high place along the way,
But in having made the journey, stage by stage,
    A sacred pilgrimage.
Birth is a beginning
And death a destination.
And life is a journey,
A sacred pilgrimage—
    To life everlasting.

With permission from The New Union Prayer Book. Central Conference of American Rabbis, New York, 1978, p. 283

Journey Across the Life Span:
Human Development and Health Promotion

Elaine Polan, RNC, MS

Assistant Supervisor
Practical Nurse Program
Vocational Education and Extension Board
School of Practical Nursing
Uniondale, New York

Daphne Taylor, RN, MS

Classroom/Clinical Instructor
Practical Nurse Program
Vocational Education and Extension Board
School of Practical Nursing
Uniondale, New York

F. A. DAVIS COMPANY • Philadelphia

F. A. Davis Company
1915 Arch Street
Philadelphia, PA 19103

Printed in the United States of America

Last digit indicates print number: 10 9 8 7 6 5 4 3 2 1

Nursing Publisher: Robert G. Martone
Nursing Editor: Alan Sorkowitz
Developmental Editor: Ann Houska
Production Editor: Jessica Howie Martin
Cover Designer: Louis J. Forgione

Library of Congress Cataloging-in-Publication Data

Polan, Elaine
Journey across the life span: human development and health promotion / Elaine Polan, Daphne Taylor.
p. cm.
Includes bibliographical references and index.
ISBN 0-8036-0196-4 (pbk.)
1. Life cycle, Human. 2. Health promotion. 3. Practical nursing.
I. Taylor, Daphne II. Title.
[DNLM: 1. Human Development—nurses' instruction. 2. Health Promotion—nurses' instruction. 3. Nursing, Practical—education.
WS 103 P762j 1998]
RT69.P65 1998
613—dc21
DNLM/DLC
for Library of Congress 97-39790
CIP

To my family: Mom, my guiding force,
and my sons Barry and Robert.
You have made *my* journey so far meaningful and full of love.

Elaine Polan

To my husband Robert, my Mom, daughter, and grandchildren for their love, support, and patience in seeing this project to completion.

Daphne Taylor

Preface

Current trends indicate a need for a change in our healthcare system to one that focuses on universal health care. This creates a need for emphasis on health promotion, maintenance, and restoration. In this new health delivery system healthcare workers are expected to provide care to individuals throughout the life span in a variety of settings. This textbook is designed to assist nursing students in their study of the life cycle from conception to old age. Instead of having to read only certain sections of a core text or portions of a pediatric or maternity text, students can now see the complete presentation of growth and development across the life span. We hope this will be meaningful and will assist them in developing an appreciation for individuals in their struggle to maintain, promote, and restore health.

There are 11 chapters, each designed to make the book user-friendly. Each chapter is preceded by a chapter outline, along with a list of learning objectives and a list of key terms. Key terms are considered important to the reader's understanding of the material. A special feature used in this text is the "Helpful Hints" box, which is designed to draw attention to important facts. Other pedagogical features include tables, illustrations, and photographs.

At the conclusion of each chapter is a summary that highlights key points, followed by a critical thinking exercise to increase awareness and challenge thinking, multiple-choice questions to help students test their understanding of the contents, and suggested readings that will enable them to further explore and research topics of interest.

We wish to point out that the names of persons used in critical thinking exercises and case studies are fictional and that any resemblance to names of actual persons is coincidental.

It is our hope that students will find this text easy to read and applicable to clinical practice and personal growth.

Elaine Polan
Daphne Taylor

Acknowledgments

We wish to extend special thanks to the following individuals, without whom we could not have completed this task.

We wish to thank Michael K. Gilroy, Executive Director of the Vocational Education and Extension Board, for supporting and encouraging our project and allowing us to use the computers and offices after hours.

Special thanks to our friend, Clifford J. Newman, who patiently taught both of us the use of the computer. Clifford made himself available for endless technical support. We are truly indebted to him for all of his assistance.

Additional thanks to Alan Sorkowitz, Developmental Editor, F. A. Davis Publishers, who always showed confidence in our ability to complete this project. Thanks to Fran Mues for her editorial suggestions and guidelines, which served as a road map for us to follow. Lastly, we owe thanks to the many students whom we have taught over the years. Their hard work and dedication have been our true inspiration.

Consultants

Donna J. Burleson, RN, MS
Director of Health Occupations
Cisco Junior College
Abilene, Texas

Barbara Cushman, RN, BSN
Southeastern Technical Institute
South Easton, Massachusetts

Evangeline Dumont, RN, MSN
Instructor
Youville Hospital School of Practical Nursing
Cambridge, Massachusetts

Sherry G. Fader, RNC, MSN
Former Chair, Practical Nursing Program
Quincy College
Quincy, Massachusetts

Corrine R. Kurzen, RN, MSN
President
Quality Continuing Education and Training, Inc.
Philadelphia, Pennsylvania

Prescilla P. LaHann, RN, BSN
Practical Nursing Coordinator
Idaho State University School of Applied Technology
Pocatello, Idaho

David K. Miller, RNC, BSN, MSEd
Indiana Vocational-Technical College
School of Practical Nursing
Columbus, Indiana

Carolyn Pearson, RN, BSN, MEd
Director of Practical Nurse Program and Supervisor of County Health Department
Bristol-Plymouth Regional Technical School
Taunton, Massachusetts

Roberta P. Ramont, RN, MS
Instructor
North Orange County ROP Vocational Nursing
Anaheim, California

Martha E. Spray, RN, MS
Adult Practical Nursing Instructor
Mid-East Ohio Vocational School
Zanesville, Ohio

Geraldine M. Wagner, MEd, BSN, RN
Director
Vocational Nursing Program
Glendale Career College
Glendale, California

Contents

Healthy Lifestyles

Chapter 1

Chapter Outline

Healthy Lifestyles

Key Words

apathy
disease prevention
empowerment
equilibrium
fight-or-flight response
general adaptation syndrome (GAS)
health
health promotion
health restoration
holistic
lifestyle
malnutrition
regression
separation anxiety
stress
wellness

Learning Objectives

At the end of this chapter, you should be able to:

- Describe the concept of health.
- List five healthy lifestyle practices.
- State the role of the practical nurse in health promotion.
- List two factors that interfere with people's ability to change their personal habits.

Today's nurse must be knowledgeable about what constitutes health, as one of the primary goals of nursing is to assist the individual in achieving the promotion of the highest level of health. In 1947, the World Health Organization (WHO) defined **health** as "a state of complete physical, mental, and social well-being, not merely the absence of disease or infirmity." The authors here attempt to define for the reader a concept of health that is **holistic** in its approach. That is, we consider health to include not only physical aspects but also psychological, social, cognitive, and environmental influences. Physical health is influenced by our genetic makeup. This includes all the characteristics that people inherit from their parents. These characteristics not only include physical features but may include genetic weaknesses or disease. In a later chapter, genetic inheritance is further explored. Psychological health refers to how a person feels and expresses emotions. Social health, on the other hand, deals with everyday issues of economics, religion, and culture as well as the interactions of people living together. Cognitive health encompasses the person's ability to learn and develop. Environmental health concerns itself with environmental issues such as air quality, noise, water, and biochemical pollutants.

Throughout this text the authors refer to specific developmental theorists to support the holistic view of growth and development. These theorists include Freud (psychoanalytic theory), Erikson (psychosocial theory), Piaget (cognitive theory), Maslow (human needs theory), and Kohlberg (moral theory). The holistic approach to health, which recognizes individuals as whole beings, promotes consideration of all aspects of a person's life. This helps the practical nurse to understand each person and attach significance, value, and meaning to each life. The holistic view further helps identify similarities and differences among people, allowing decision making from the person's own unique perspective. Positive nursing outcomes using the holistic approach emphasize patient independence and maximize potential.

Throughout this text we use the terms *health* and **wellness** synonymously. We believe that health, from the holistic perspective, is a balance of internal and external forces that leads to optimal functioning (Table 1–1). True health produces a state in which individuals are able to meet their needs and interact with their environment in a mutually beneficial manner. Healthy individuals exhibit effective coping patterns and experience a certain degree of comfort and pleasure in their activities. Health may be visualized on a scale or continuum (Fig. 1–1). One end of the continuum depicts optimal health or wellness, whereas the other end shows disease, total disability, or death. Disease refers to an imbalance between the internal and external forces. Individuals find that, throughout the life cycle, health

TABLE 1–1

A HOLISTIC MODEL OF HEALTH

Internal forces	Body systems Mind Neurochemistry Heredity
External forces	Culture Community Family Biosphere

Health-Wellness	Disease	Total Disability	Death

FIGURE 1–1
The health-illness continuum.

is not static but dynamic and can move backward and forward from a state of wellness to illness or disease.

Traditionally, health care has focused on an illness model, in which the primary role of the nurse is to relieve pain and suffering. Today disease prevention is evolving as an area of nursing concern. This change places new demands on the practical nurse, with a greater emphasis on his or her role in patient education and health promotion through all stages of the life cycle.

PROMOTING, MAINTAINING, AND RESTORING HEALTH

Health promotion means health care directed toward the goal of increasing one's optimum level of wellness. Healthy life means full functional capacity at each stage of the life cycle, from infancy through old age (Fig. 1–2). Promotion of health can occur at any time and is related to individual lifestyles and personal choices. Health promotion allows people to enter into satisfying relationships at work and play. Health means being vital, productive, and creative and having the capacity to contribute to society.

FIGURE 1–2
A healthy lifestyle helps people to reach full functional capacity at each stage of the life cycle.

The national aspirations for health promotion include three goals: healthy lives for more Americans, elimination of healthcare disparities among all ethnic and racial groups, and access to preventive services for everyone. The essential component of health promotion begins with the sharing of knowledge. The acquisition of knowledge then influences attitudes and leads to a change in behavior. Health promotion is most successful when placed in a supportive social environment. This environment first begins within the home and extends into the community. The community includes schools, churches, and businesses. Schools provide the location for the dissemination of health information among the young. More than 85 percent of American adults spend the greater part of the day in the workplace. The workplace, therefore, is another excellent site to continue educating adults on health issues. Health promotion emphasizes nutrition, exercise, mental health, and avoidance of substance abuse. These health promotion issues are addressed throughout the text as they relate to specific age groups.

The practical nurse's first step in promoting health is assessing individuals and families for potential risks. Physical, social, and individual values must be considered essential components of the nursing assessment. Nurses can encourage patients to assume full responsibility for their behavior and to adopt a healthier **lifestyle**. **Empowerment** is a form of self-responsibility in that it demands that people take charge of their own decision making. The practical nurse can play an important role in educating and guiding patients so that they have enough information to make critical decisions and be informed health consumers. Nurses must develop ways to help patients recognize their own needs, solve their own problems, and access resources that give them a sense of control over their own lives. Acts that empower patients place the nurse in the role of patient advocate. Nurses must be careful not to express their own personal opinions but to share enough information that the patients can make informed decisions. An example of the nurse's role as an empowering agent occurs when a newly diagnosed cancer patient attempts to choose among the different treatment modalities offered by the physician. The nurse helps to assess the patient's knowledge level, communicates clear information, and supports the patient's decision concerning treatment. Throughout this text we discuss health-promoting activities that enable patients to maintain wellness, strive for their full potential, and enjoy a high quality of life.

Disease prevention is composed of three levels: primary, secondary, and tertiary. Primary prevention occurs before there is any disease or dysfunction. The term "health promotion" may sometimes be used interchangeably with "primary prevention." Examples of primary prevention include patient education on basic hygiene, nutrition, and exercise. Other examples of primary prevention may include immunizations against infectious diseases, avoidance of substance abuse, and regular dental examinations.

The most significant public health achievement in the last century is the reduction in the incidence of infectious diseases. Many factors have helped eradicate infectious diseases. Improvements in basic hygiene, food handling, and water treatment and the widespread use of vaccines have contributed to disease control. Antibiotics have further helped to successfully treat infectious diseases. Infants, older adults, minority groups, and healthcare workers are at increased risk for infectious diseases. All causative organisms, even those that are presently rare, may pose a potential threat of recurrence. One major national goal is to effectively deliver immunizations to at least 90 percent of the preschool population. Special efforts should target minority populations, particularly African Americans and

Hispanics, who continue to have lower immunization levels than the general population.

To some extent, certain environmental hazards, such as air, water, noise, and chemical pollution, are beyond our control. But we are all still responsible for becoming informed and must take an active part in limiting the effects of these hazards. In addition, personal injury is one of the leading causes of death in the United States. We must each identify and reduce our high-risk behaviors at home, on the job, or when we travel. The different health hazards that exist at each stage of the life cycle are outlined in later chapters.

Secondary prevention begins with diagnosis of disease or infectious processes. It focuses on the need for early diagnosis and treatment of disease to prevent permanent disability. This includes all interventions used to halt the progress of an already existing disease state. Secondary prevention includes screening of all types (e.g., breast self-examination, testing for hypertension and sickle-cell disease). In the latter stages of disease, secondary prevention often includes activities that prevent further disability. For example, the practical nurse may help the diabetic patient prevent ulceration and the loss of a limb by instructing and practicing good foot care.

Tertiary prevention begins when a permanent disability occurs. Tertiary prevention is also referred to as **health restoration**. Health restoration begins once the disease process is stabilized. Nursing care is directed toward rehabilitation and restoring the person to an optimal level of functioning. The goal of tertiary prevention is to regain lost function and develop new, compensatory skills, possibly with the use of assistive devices such as a cane or hearing aid. Another goal is to help patients, including those with incurable diseases, attain the maximum level of health. In helping patients achieve this objective, the nurse may collaborate with other health professionals, such as physical and occupational therapists. The nurse is also responsible for offering psychological support to patients and family members. The practical nurse can help an elderly stroke patient to achieve optimal functioning by first treating the patient with dignity and respect. Secondly, the nurse must provide individualized care that maximizes the person's strengths and minimizes weaknesses.

HEALTHY LIFESTYLES

In recent years it has become evident that many illnesses are preventable and that certain lifestyles greatly reduce the incidence of heart disease, stroke, and other diseases. There is a strong need to emphasize healthy behavior to reduce our current mortality rates. The overwhelming need is to learn to identify behaviors that place individuals at great risk of acquiring and possibly spreading disease. Health promotion campaigns should target the role of nutrition, exercise, mental health, substance abuse, self-concept, and disease prevention (Box 1–1). Important to the success of any health promotion plan is the need to motivate and encourage patients to take responsibility for their own actions and healthcare practices.

Nutrition

Nutrition is an important factor in promoting optimum health. Studies have shown that the five leading causes of death associated with poor dietary habits are coronary

BOX 1–1

Health-Promoting Behaviors

Sound nutritional practice
Regular physical exercise
Stress management
Chemical avoidance
Disease prevention
Healthy self-concept

heart disease, cancer, stroke, noninsulin diabetes mellitus, and coronary artery disease. **Malnutrition** is poor dietary practice that results from the lack of essential nutrients or the failure to use available foods. Malnutrition may involve undernutrition, including symptoms of deficiency diseases, or overnutrition. This text focuses on dietary needs at each stage of growth and development (Fig. 1–3).

An adequate diet provides sufficient energy, essential fatty acids and amino acids, vitamins, and minerals needed to support optimum growth and maintain and repair body tissues. It is not possible to design one diet for everyone, as individual needs for nutrients vary greatly with age, sex, growth rate, and amount of physical activity as well as other factors. Because most nutrients are widely distributed in a variety of foods, it is very possible to design a diet or meal plan that satisfies an individual's personal and cultural preferences and lifestyle needs.

FIGURE 1–3
A healthy lifestyle includes exercise and good nutrition.

Exercise

Regular exercise improves muscle strength and endurance, increases lung capacity, decreases tension and stress, and helps maintain adequate cardiovascular functioning. Studies have confirmed that regular exercise is useful in preventing heart disease, osteoporosis, and other illnesses. The optimum benefits of exercise are seen when physical fitness is maintained throughout the life span (Fig. 1–4). Table 1–2 lists the positive effects of exercise.

There are two important points to remember about exercise: First, before beginning an exercise program, the individual should check with a physician. Second, moderation is better than excessive practice. The need for exercise and its applicability to the stages of growth and development are explored in each chapter of the text.

Mental Health

Mental health is a fluctuating state in which the individual attempts to adjust to new situations, handle personal problems without undue stress, and still contribute to

FIGURE 1–4
Walking is an excellent form of exercise to promote and maintain health.

TABLE 1–2
POSITIVE EFFECTS OF EXERCISE

Cardiovascular system	Increases blood volume and oxygen content Increases blood supply to muscles and nerves Decreases serum triglycerides and cholesterol levels Reduces resting heart rate Increases heart muscle size
Respiratory system	Increases blood supply Increases exchange of oxygen and carbon dioxide Increases functional capacity
Neurological system	Reduces stress and increases euphoric effect
Musculoskeletal system	Increases muscle mass Increases muscle tone Improves posture

society in a meaningful manner. Mentally healthy individuals see themselves and others realistically. The state of mental health fluctuates from day to day but maintains a certain degree of continuity and consistency. Certain behaviors may be normal in moderation but unhealthy in excess. For example, washing one's hands as part of everyday hygienic practices is considered acceptable. However, repeated handwashing unrelated to any activity is seen as bizarre and mentally unhealthy. A factor that may affect one's mental health is stress. **Stress** may be defined as anything that upsets our psychological or physiological **equilibrium**, or balance. Responses to stress may be physiological, emotional, or intellectual. Some of the common physiological responses to stress include an increase in heart rate, respiratory rate, and blood pressure. Emotional responses to stress include irritability, restlessness, and a sense of discomfort. Intellectual responses to stress often include forgetfulness, preoccupation, and altered concentration.

Many years ago Hans Selye described three distinct stages of physiological response to stress. This is known as the **general adaptation syndrome (GAS)**.

1. *Alarm stage:* Hormones from the adrenal cortex place the body in a state of readiness known as the **fight-or-flight response**.
2. *State of resistance:* The body attempts to adapt to the stressors.
3. *State of exhaustion:* After prolonged exposure to stress, the body's energy becomes depleted. This may result in disease or destruction.

Stress is a necessary part of life. It is unrealistic to expect to be able to eliminate all the stress from one's life, and it would not be healthy to do so. A certain degree of stress stimulates us to act and mobilizes our coping abilities. Each of us perceives stressors differently, depending on our learned behavior, age, and personality. Some individuals find air travel pleasurable, whereas others find it stressful. Stress may come from internal sources, such as illness. It may also come from external sources, such as family, school, or peers. Stress can be acute or chronic in nature. Typical stressors in childhood include such events as birth of a sibling, family death, onset of schooling, illness, and **separation anxiety**, to name a few. The practical nurse may witness stress or separation anxiety in a young child as a result of hospitalization. Separation anxiety may be seen in three phases: protest, despair, and detachment. Protest is evidenced by loud crying, restlessness, and dissatisfaction with substitute caregivers. Despair produces a sense of hopelessness and is seen as a quieter period.

Detachment is a state of withdrawal and **apathy**, or lack of interest in one's surroundings. The nurse's awareness of these stages of separation helps both the child and the caregivers cope and adapt to the stress. **Regression**, the return to an earlier stage of development, may be another childhood adaptation to stress. The stress of serious illness or hospitalization may cause the youngster to show regressive behaviors. The practical nurse can reassure the parent that bed-wetting after illness in a previously toilet-trained child is only temporary. After recovery from the illness, the child will return to the previous level of accomplishment.

Typical stressors during adolescence relate to individuals' search for their identity. Decision making and the struggle for independence lead to family discord. Nurses can best assist adolescents in their struggle for independence by being supportive and encouraging decision making.

The main stressors identified for adults are related to their key relationships. According to the Social Readjustment Rating Scale (SRRS), death of a spouse, divorce, and separation are the events perceived as the most stressful for adults. During old age, life stresses include the loss of a spouse, retirement, and illness or loss of function. Nurses can help older adults cope effectively with stress by identifying and mobilizing their available support systems. For example, when an elderly patient is hospitalized, the family members need to be involved in the plan of care. Family involvement and frequent family contact help reduce stress for all involved persons. Another part of stress management is the identification of what is perceived to be overly stressful. Following this, we must either reduce the stress or learn how to manage it in a healthy way. Healthy ways of adapting to stress include relaxation, exercise, humor, and guided imagery. Box 1–2 describes two techniques for reducing stress.

BOX 1–2

Stress Reduction Techniques

Practice these techniques several times a day.

RELAXATION EXERCISE

1. Assume a comfortable position.
2. Close your eyes.
3. Regulate breathing pattern and focus on inhaling and exhaling.
4. Progressively relax your muscles.
5. Eliminate other distractions.
6. Refocus on breathing, as needed.
7. Continue for 10 to 15 minutes.

GUIDED IMAGERY

1. Take a relaxed position.
2. Close your eyes.
3. Recall an image, event, or place that is pleasurable.
4. Focus energy and thought on the image while relaxing muscles from head to toe.
5. Concentrate on the image for 10 to 15 minutes.

Unhealthy or maladaptive responses to stress include denial, withdrawal, and acting-out behaviors. One example of denial is seen when a surviving spouse, after a prolonged period of mourning, is unable to change the deceased spouse's bedroom, leaving clothes and belongings about as if the person were going to return.

Substance Abuse

Substance abuse commonly refers to the abuse of drugs such as alcohol, nicotine, and caffeine as well as legal and illegal pharmaceutical preparations. A substance abuser has more than a strong desire for something. Substance abuse is marked by gradual reduction in awareness, decline in self-esteem, and withdrawal from involvement. Recently much attention has been given to educating the public about the effects of substance abuse. Tobacco use is one of the most important preventable causes of death in our society today. The connection between smoking and lung disease has been well documented for over 30 years. Smoking also contributes to heart disease and fetal and neonatal abnormalities and deaths. Health promotion is aimed at encouraging cessation of smoking and avoiding exposure to secondary smoke. Since all behavior is influenced by role modeling, social education and individual responsibility are needed to further reduce the incidence of smoking in future generations.

Although moderate alcohol use may be beneficial in lowering blood cholesterol levels, alcohol abuse presents a serious health problem in our society. Alcohol abuse is not measured according to the specific amount consumed; rather, alcohol abuse occurs when the person cannot curtail the amount of alcohol he or she consumes or when alcohol is interfering with the person's daily functioning. Long-term alcohol use has been linked to liver damage, heart disease, and an increased risk of neonatal disorders. Statistics indicate that alcohol contributes to serious social problems, including automobile accidents. Alcohol use is implicated in almost half of the deaths caused by motor vehicle crashes. It is also responsible for one-third of all homicides, drownings, and boating deaths.

Intravenous drug users and their sexual partners are at increased risk for acquired immunodeficiency syndrome (AIDS) and related sexually transmitted diseases (STDs). Drug treatment programs, counseling, education, and frequent testing reduce the many risks associated with drug use.

A Healthy Self-Concept

Even knowing and practicing healthy behaviors and avoiding risky behaviors do not guarantee good health. Many factors influence an individual's health status, including relationship to family and community, how the person perceives various social pressures, and individual temperament. All of these factors cause each person to respond to the environment in a unique and sometimes unpredictable manner. Furthermore, perceptions of the environment, reactions to it, and individual needs are affected by a person's self-concept.

The relationship between the individual and the environment is reciprocal; that is, a person's self-concept is affected by the environment in which he or she lives. Self-concept is affected by the individual's stage of development as well. Throughout this text we discuss the stages of development and how they might affect the individual's health and choices related to healthy lifestyle.

Chapter 3 of this text provides an explanation of several theories related to individual development: psychoanalytical, psychosocial, cognitive, moral, and so forth. Good health is more than a visit to the doctor. Healthy decisions, such as what we eat for lunch or the amount of exercise we get, can be made every day. Although many Americans are moving toward healthy living, others still need to be encouraged to practice healthy behaviors. Each person must set realistic goals that will reinforce positive behaviors. The first step is to make an inventory of healthy and unhealthy behaviors and develop a health assessment plan (Boxes 1–3 and 1–4).

BOX 1–3

Health Behavior Inventory

Before beginning any lifestyle change, it is important to assess one's behavior. List below the healthy and unhealthy things that you do.

HEALTHY BEHAVIOR

UNHEALTHY BEHAVIOR

BOX 1–4

Personal Health Assessment Plan

In any health assessment it is important to first become aware of your personal quality of life. Answer the following questions and determine you personal health action plan.

1. Based on my readings, my personal health issues of most concern are:
   ____
   ____
2. My top-priority health issue is: ______________________________
   ____
3. I can address the preceding issues by taking the following actions: ______
   ____
   ____
4. My goal is to: ______________________________
   ____

ROLE OF THE NURSE IN HEALTH PROMOTION

As changes in the healthcare system take place, the role of the practical nurse also needs to change to meet the demands of the new healthcare delivery system. Practical nurses will need to work not only in traditional hospital settings but also in the community. Emphasis will be on prevention and health promotion. All levels of nurses will be accountable to both the employer and the client.

In the future, the practical nurse will have five roles and responsibilities.

1. *Caregiver:* delivering healthcare services
2. *Teacher:* educating the client, family, and community
3. *Advocate:* helping clients choose between available options
4. *Collaborator:* working as member of a team, sharing and exchanging information
5. *Role model:* practicing healthy lifestyle behaviors that will influence and reinforce clients' actions

---

SUMMARY

1. Health is a balance of internal and external forces leading to optimal functioning.
2. Health promotion refers to health care directed toward increasing one's optimal level of wellness.
3. A change in lifestyle or personal habits is often necessary to promote maximum health.
4. Health maintenance focuses on prevention and the need for early diagnosis and treatment.
5. Health restoration begins after the disease process is stabilized. The goal is to either restore function or help the person compensate for losses.
6. Healthy lifestyle includes attention to nutrition, exercise, mental health, substance abuse avoidance, and disease prevention.
7. Factors that influence a person's health behavior include family, role models, social pressures, and one's self-concept.
8. Disease prevention is comprised of three levels: primary, secondary, and tertiary.
9. The five roles for the practical nurse in health promotion are caregiver, teacher, advocate, collaborator, and role model.

CRITICAL THINKING

Larry Woodhill, a 47-year-old bank manager, attends the wellness clinic offered by his organization. Larry weighs 230 pounds and is 5 feet, 8 inches tall. He has smoked two packs of cigarettes per week for the past 30 years. Recreational activities begin on Friday nights at the local bar, where Larry consumes five or six cans of beer. This activity is repeated on Saturday nights, but not on Sundays. Through the wellness clinic, Larry embarks on an exercise program.

1. Without other major behavior changes, how do you evaluate the benefits of exercise for Larry?
2. Develop an approach to dietary counseling that may be beneficial for Larry.
3. What other healthy choices could Larry incorporate into his lifestyle?
4. Which of Larry's behaviors are considered high-risk?

Multiple-Choice Questions

1. Health can be defined as:
   a. Harmony between illness and wellness
   b. Balance between internal and external forces
   c. A state of mental thought process
   d. A state of physical functioning

2. The objective of health promotion is:
   a. To hold the professional nurse responsible for the client's lifestyle practices
   b. To provide positive reinforcement to secure each healthy act
   c. To achieve an optimum level of wellness
   d. To decrease the person's tolerance level for stressful events

3. Empowerment is:
   a. The intensity of feelings generated by the client
   b. A needs-based behavior
   c. Client-centered decision making
   d. Open interaction between the client and the environment

Suggested Readings

Anspaugh, DJ, Hamrick, MH, and Rosato, FD: Wellness: Concepts and Applications. Mosby-Year Book, St. Louis, 1991.
Bigbee, JL, and Jansa, J: Strategies for promoting health protection. Nurs Clin North Am 26(4):895–1001, 1991.
Brehm, BA: Essays on Wellness. HarperCollins, New York, 1993.
Dixon, JK, Dixon, JP, and Hickey, M: Energy as a central factor in the self-assessment of health. Advanced Nursing Science 15(4):1–12, 1993.

Edelman, CL, and Mandle, CL: Health Promotion throughout the Lifespan. Mosby-Year Book, St. Louis, 1990.
Gillis, A: Determinants of a health-promoting lifestyle: An integrative review. J Adv Nurs 18:345–353, 1993.
Green, J, and Shellenberger, R: The Dynamics of Health and Wellness: A Biopsycho-Social Approach. Holt, Rinehart & Winston, Fort Worth, TX, 1991.
Health United States: Healthy People 2000 Review. Department of Health and Human Services Pub. No. (PHS) 93-1232. Hyattville, MD, 1992.
Igoe, JB, and Giodano, BP: Health promotion and disease prevention: Secrets of success. Pediatric Nursing 18(1):61–66, 1992.
Jones, PS, and Meleis, A: Health is empowerment. Advanced Nursing Science 15(3):1–14, 1993.
Kelley, MP: Health promotion in primary care: Taking account of the patient's point of view. J Adv Nurs 17:1291–1296, 1992.
Knollmueller, RN: Prevention across the Life Span: Healthy People for the Twenty-First Century. American Nurses Publishing, Washington, DC, 1993.
Lusk Lechlitner, S., Kerr, MJ, and Ronis, DL: Health-promoting lifestyles of blue-collar, skilled-trade, and white-collar workers. Nurs Res 44(1):20–24, 1995.
Mayhew, MS: Strategies for promoting safety and preventing injury. Nurs Clin North Am 26(4):885–892, 1991.
Redland, AR, and Stuifbergen, AK: Strategies for maintenance of health-promoting behaviors. Advances in Clinical Nursing Research 28(2):427–441.
Spellbring, AM: Nursing's role in health promotion: An overview. Nurs Clin North Am 26(4):805–813, 1991.
Wilhide Tanner, EK: Assessment of a health-promotive lifestyle. Nurs Clin North Am 26(4):845–853, 1991.

The Family

Chapter 2

Chapter Outline

The Family

Key Words

autonomy
culture
dysfunctional family
functional family
infant mortality rate
nurturance
omnipotence
socializing agent

Learning Objectives

At the end of this chapter, you should be able to:

- Give the classic definition of the term "family."
- Describe the eight family types.
- Name two groups that assist the family in socializing the child.
- List the four different stages of family development.
- Contrast the characteristics of functional and dysfunctional families.

The family unit is the place where the individual first learns to make decisions that will enable the promotion of health and well-being. The family is where both children and adults are loved, protected, and taught. Individuals learn about themselves, their relationships, and their behavior within the family unit (Fig. 2–1). Each person in the family unit plays a role in the other members' health picture. Changes in the state of one member's health or illness may affect other family members. The nurse's understanding of the importance of the family helps to provide rationales and guidelines for clinical practice.

Healthcare workers must recognize the patient as a part of a family unit, not in isolation. This holistic approach to health care requires that the licensed practical nurse (LPN) be familiar with the meaning of today's family: its functions, types, stages, size, patterns, and cultural issues. Knowledge of culture and ethnicity will help the nurse to better understand how these issues affect a person's health actions and practices. In addition, it is important that nurses be not only aware of different family variations but open and nonjudgmental in their approach to patient care.

Until fairly recently, the basic family unit has usually been defined as two or more people related by blood, marriage, or adoption who live together. This definition of "family" is narrow in its scope and does not accommodate the many different living arrangements that are in place today. A more current definition of today's family might be "two or more people who have chosen to live together and share their interests, roles, and resources." Each family is unique in its style and makeup, but usually attachment and commitment are the features that bind people together.

FAMILY FUNCTIONS

The family is one of the most important and powerful groups that individuals belong to. Although each family is set up for a specific purpose, a common goal shared by all families is the growth and development of its members. The family progresses

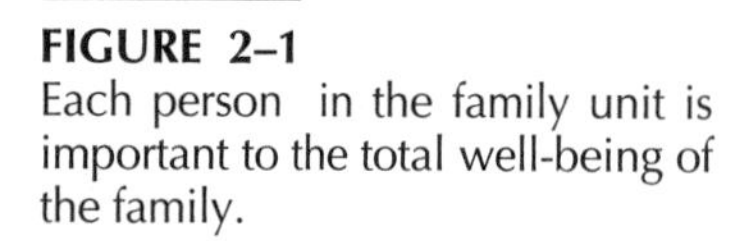

FIGURE 2–1
Each person in the family unit is important to the total well-being of the family.

through distinct stages of development over time, but the ultimate goal of the family is the survival and personal fulfillment of each member. Each family has certain distinct strengths and weaknesses, but all families share certain basic characteristics.

All families have a basic purpose or function and a set of values and governing rules. Several functions help secure this goal. These functions are not exclusive to families alone, but in combination they are unique to this institution. The basic functions of the family include physical maintenance, protection, nurturance, socialization and education, reproduction, and recreation.

Physical Maintenance

The family must provide food, clothing, water, and shelter for each of its members. The ease and the manner in which they can provide these necessities vary with each family and depend on the unit's economic success. Nearly one out of every eight American families lives below the federal poverty level. This low socioeconomic status results in a health disparity between the poor and better-income families. Public assistance programs have been set up to help needy families meet basic family needs.

Protection

Each family member needs protection against inherited and acquired illnesses (internal forces) and injury (external forces). Protection may take different forms at different points of the life cycle. Before and during pregnancy, health screening and genetic counseling may offer individuals protection against certain diseases, including inherited diseases. After birth, immunization protects the infant and child against a number of illnesses. (The recommended schedule is listed in Appendix B.) Diet, exercise, and health screening help protect adults from illness. Individuals in families are protected from external forces such as injury throughout all stages of growth and development. This is best accomplished through education, awareness training, and role modeling.

Nurturance

The family provides **nurturance**—loving care and attention—to each of its members. In fact, it is the only group that offers almost unconditional acceptance, love, and emotional support for its members. Young infants need touching, cuddling, and the sound of the caregiver's voice as well as food if they are to thrive. As they grow, children also need to have limits set on their behavior. Without such discipline, a child feels unprotected and unloved. Adult family members need to nurture and care for each other as the need for love continues throughout the life cycle. If the family unit breaks down, other support systems must fill this need. Illness or death of a family member may result in stress on the family unit. Divorce is another common example of family breakdown. In families affected by divorce, nurturance can continue with support from the extended family or referrals to community support groups.

Helpful Hints: Understanding a Child's Concept of Death

Children have different concepts of death at different ages. Their responses reflect their stage of emotional and cognitive development.

- Infants have no concept of death.
- Toddlers believe that death is temporary or reversible.
- Preschool children believe that their thoughts may cause death. This causes feelings of guilt and shame.
- School-age children understand the permanence of death but may associate it with misdeeds. They sometimes personify death as a monster or other evil thing.
- Adolescents have a mature understanding of death but may be subject to guilt and shame. This age group is least likely to accept death, especially if it happens to one of their peers.

Socialization and Education

The family is the child's primary **socializing agent**: Children first learn how to interact with their social environment by observing how other family members act and respond (Fig. 2–2). It is within the family that the child first learns about the world and how to respond to it. The education of the child begins in the home. Other important socializing institutions—notably, the schools—may support and supplement the family unit, but educational success cannot be accomplished unless both family and school work together toward a common goal.

Today, there is much debate about whether the schools can—or should—teach certain values and provide information about topics that have traditionally been considered part of the family's domain, for example, sex education and drug awareness programs. Other institutions and agents—the church, the media, or organizations

FIGURE 2–2
The family is the primary socializing agent for children.

such as the Scouts—may support and supplement the family, but the family unit is still the primary socializing institution.

Reproduction

Reproduction is the means by which the family survives and passes its genes to succeeding generations. Reproduction is a bodily function that begins with puberty. However, it requires not only physical readiness but psychological preparedness and a lifetime of commitment.

Recreation

The family unit should be able to spend time together in pleasurable activity. It is important to the success and cohesiveness of the group that family members share fun time as well as work and other roles. This creates a balance and opens the channels for communication. In today's family, "free time" may be difficult when both parents work or a single parent must play multiple roles.

FAMILY TYPES

In today's world, there are a number of different family structures or types. The most common types include the nuclear or conjugal, the extended, the single-parent, the blended or reconstituted, the cohabitative, the communal, the foster or adoptive, and the gay or lesbian family. For a summary of family types, see Table 2–1.

The Nuclear or Conjugal Family

The nuclear or conjugal family, also known as the traditional family, consists of a husband, a wife, and their children. Today statistics indicate that fewer than one-third of all families are of the nuclear type. Families consisting of two adults without children are referred to as a nuclear dyad. Marriage is the main binding force in both of these family types. In recent history, the nuclear family has become the model for

TABLE 2–1
FAMILY TYPES

Types	Members
Nuclear, conjugal, or dyad	Husband and wife, with or without children
Extended	Husband and wife, children, and grandparents
Single-parent	Mother or father and children
Blended or reconstituted	Mother or father, stepparent, and children
Cohabitative	Man, woman, and children
Communal	Individuals with their mates and children
Foster or adoptive	Parents or caregivers and children
Lesbian or gay	Two women or two men, with or without children

other more complex family types. Traditionally, the man was expected to be the breadwinner, and the woman was considered the caregiver and homemaker. In today's nuclear family, both parents are probably in the workforce and there is a sharing of roles. More and more fathers are actively involved in raising their children. Someone other than the parents, or some outside agency, may provide child care while the parents are at work.

The Extended Family

The extended family type consists of the nuclear family plus grandparents, aunts, uncles, or cousins living together under the same roof. Children in this household have many models from which to choose. In this family type there may be a sharing of resources and roles. Elders living in the extended family can assist with child-rearing roles. This may strengthen their need for usefulness and increase their sense of belonging. The undesirable effect of the extended family occurs when the older person is undervalued and seen as a burden. It is important that nurses in all practice settings be able to assess and evaluate the coping abilities of extended families.

During the latter part of the twentieth century, the extended family has gradually returned and become more commonplace as a result of certain outside forces—namely, increases in the cost of living, unemployment, longer life spans, and greater numbers of divorces and teenage parents. The extended family may provide a temporary respite from economic or social hardship; once recovery is achieved, family members may move out on their own. Often, however, the nuclear family may be set up in close proximity to parents and other relatives. In this case, the nuclear family has some of the feel of an extended family because of the regular, frequent contact between family members.

The Single-Parent Family

The single-parent family consists of an adult living with one or more children. In most cases, single parents are divorced, separated, or widowed. However, a growing number of adults are choosing this family type as an alternative lifestyle. Today, 60 to 70 percent of families are single-parent families. Most are headed by women, although recently, more men are becoming single parents. A major challenge of this family type is that the single parent must assume the role of both caregiver and breadwinner. Single parents may look to their own family of origin for support and assistance. Other outside agencies or individuals may also assist this family type.

There is current evidence that the marital status of a child's parents will affect the child's health status. In 1983, unmarried motherhood was associated with a higher infant mortality rate for both blacks and whites. The **infant mortality rate** is the number of infant deaths per 1000 live births in the first year of life. The mortality rate was 13.1 per 1000 live births for married white women and 14.1 per 1000 live births for married black women. The infant mortality rate for unmarried black women is 19.6 per 1000 live births.

Divorce or separation may increase health risks for children. Recently a study indicated that the children of divorced or separated parents were at one-third greater risk of developing problems, including pneumonia, ear infections, and tonsillitis, than children from intact families.

The Blended or Reconstituted Family

The blended or reconstituted family is created when one or both partners bring children from a previous marriage into the relationship. Divided loyalties and resentment toward the stepparent can create stresses, which may be compounded if one parent must pay support for a child living in another household. In addition, children have to adjust to multiple views, attitudes, and personalities. Conflicts frequently emerge over how to discipline the children. Open communication between family members is essential in resolving conflicts and uniting all the parties. After the initial adjustment period, the members may unite to form a new, congenial group.

Helpful Hints: Parenting Stepchildren

- Share and value history and memories through stories, pictures, and videos.
- Encourage respect for individual differences.
- Give everyone a place for belongings.
- Avoid taking sides and showing favoritism.
- Establish a united approach to child care.
- Avoid negative comments about the absent parent.
- Be sensitive to children's concerns about differences in their surnames.

The Cohabitative Family

In the cohabitative family, a man and woman choose to live together without the legal bonds of matrimony, but in all other ways this type of family resembles the nuclear or blended family. Recently, this family type has gained popularity before or between marriages. Many of these families include children from previous relationships. These relationships may be less stable and are subject to change at any time. Stability increases when couples remain together for a long period of time.

The Communal Family

The communal family consists of a group of people who have a common philosophy, value system, and goals and choose to live together, sharing roles and resources. All the children become the collective responsibility of the adult family members. This family style became popular in the 1960s as a result of the political ferment of the period and disenchantment with society. It is difficult to track and document but still exists in rural areas.

The Foster or Adoptive Family

Foster families are those that take temporary responsibility for raising a child other than their own. This is usually a temporary placement, but it may extend over a long period depending on the stability of the birth family. This type of family faces a number of challenges. If the foster child is from a dysfunctional family, the child may experience behavioral problems as he or she attempts to cope in the new environment. The age of the child and the length of time that he or she has been in foster care will affect the child's ability to make the transition in the new setting. Foster parents assume legal responsibility for the child in their care.

The adoptive family permanently adds a child other than its own to its structure. This child has all the legal entitlements of a birth child. Adults who choose to adopt may do so because they want children but cannot or do not wish to give birth to a child. In the past, adoption records were not made public. Today many adoptees seek out their birth parents in order to better understand their identity and personal history.

The Gay or Lesbian Family

Gay or lesbian families can also take the form of any of the preceding families except that they consist of two adults of the same sex living together and sharing common emotional bonds, resources, and parenting roles. Society's attitudes toward gay and lesbian relationships have become somewhat more liberal in recent decades. The courts are increasingly willing to award child custody to homosexual parents and to allow the nonbiological parent in a homosexual couple to adopt the partner's child. Gay and lesbian couples sometimes choose to adopt as a means of meeting their nurturing needs.

FAMILY STAGES

The following is a brief description of family stages. Not all families go through every stage. For example, a couple without children may still be considered a family. Other families may not survive into old age.

Couple Stage

Traditionally, a new family is launched when young single adults decide to move away from their family of origin and start a unit by themselves. When two people form an affectionate bond and move in together, they become a couple (Fig. 2–3). This is the first stage of a new family, and emotionally it may be quite difficult as each partner merges his or her original values and beliefs with those of the new partner. Many adjustments are necessary as each partner learns to accept the other's habits, preferences, and routines. Also, early in the couple stage, the couple will need to define roles and distribute and accept responsibilities. This is a very important move. It allows the young person to try out the roles and values learned

FIGURE 2–3
Affection and bonding begin the couple stage.

in his or her family of origin and test newly acquired skills and independence. This experience can be both exciting and threatening to the young adult. Throughout this testing period, the individual may remain tied to the family of origin. During stressful times the young adult may rely heavily on the family of origin for financial and emotional support.

One of the objectives of this union is to establish a satisfying relationship built on mutual respect. Both parties must be able to compromise and to recognize and accept the other person's point of view. Sometimes this means putting aside one's own needs and considering the needs of the other person.

Sometimes a couple decides to postpone marriage until their careers are started. Postponement may have certain advantages and disadvantages. These individuals may be more mature but also more set in their ways. Communication channels must be kept open to maintain a healthy, satisfying relationship. Intimacy must be valued but not to the exclusion of each partner's **autonomy**, or independence and sense of self. Pleasurable activities, humor, and relaxation should be integrated into the couple's daily living. It is important to the success of the marriage or relationship that the couple be separate from but still closely connected to their families of origin (Box 2–1).

Childbearing Stage

The arrival of a baby changes the family constellation dramatically. Both parents must have time to adjust to their new and expanded roles. Early preparation for parenthood can help decrease some of the anxiety and stress for the new parents.

When making decisions about child care, the mother and father should each consider the other's philosophy. Care and development of the child and parents are also enhanced by close interactions with grandparents and other relatives. Even with a close relationship to extended family members and with expansion of the family as other children are born, each family member must make new role adjustments

BOX 2–1

Maintaining a Healthy Relationship

1. Clarify roles with families of origin while maintaining self-identity.
2. Permit autonomy while reaching out to maintain intimacy.
3. Value time for privacy.
4. Recognize and seek support from outside agencies during periods of stress.
5. Tighten family bonds in times of stress or crisis.
6. Respect other members' worth.
7. Handle anger and conflicts with open communication.
8. Maintain a sense of humor.
9. Satisfy your mate's need for security and safety.
10. Demonstrate caring while maintaining a romantic outlook.
11. Be open and tolerant to the other person's point of view.
12. Take time to have fun and share with each other.

without compromising his or her autonomy and sense of self. Parents and children alike can thus develop confidence and enhance their self-worth.

Grown-Child Stage

Once again the family must make adjustments to the new family unit. Grown children leave home and start on their own. This is sometimes described as the "empty-nest syndrome." The parents now shift their focus from caring for the children to caring for each other once again. This can also be a time for the development of new roles, interests, and accomplishments. Many adults return to school or begin new careers during this stage. This may be a very rewarding period, allowing each partner to fulfill lifetime goals.

Older-Family Stage

The transition into the elderly years generally begins with the retirement of one or both spouses. Perceptions of retirement are often based on economic preparation and physical health. Many elderly families prefer to live separate from but within close proximity of their children. Older adults must often make several adjustments because of changing health, declining income, and reduced energy. Some older adults must also adjust to the death of a spouse and the resultant role changes that occur at this point in life. Older adults should be sure to include pleasurable recreational activities in their daily lives. Many continue to maintain rich, rewarding relationships with their children and grandchildren throughout their older years. These kinds of pleasurable activities help the older person maintain a high level of self-esteem. (See Table 2–2 for a summary of family stages.)

TABLE 2–2
FAMILY STAGES

Stage	Task
Couple stage	Establish bonds between individuals
	Adjust to new routines
	Define roles and responsibilities
Childbearing stage	Integrate baby into the family unit
	Adjust to new roles, extend relations to extended family
	Explore and establish child-care philosophy
Grown-child stage	Adjust to new roles, empty nest
	Focus on reestablishing marital relationship
	Develop new roles, interests, and accomplishments
Older-family stage	Adjust to retirement living
	Adjust to decline in income
	Adjust to changing health and reduced energy
	Maintain rewarding relationships with children and grandchildren
	Establish pleasurable activities to build self-esteem

FAMILY SIZE, BIRTH ORDER, AND GENDER OF CHILDREN

Decisions about family size are very important. Family planning, the spacing and numbering of children in the family, requires both maturity and responsibility. Effective family planning or the avoidance of unwanted pregnancy can improve overall infant health. It has been shown that women who plan their pregnancies tend to seek out earlier prenatal care than those who have unplanned pregnancies.

The family unit is not constant; it changes with the addition of each new member. Each child has a distinct place in the family. A child's birth order can provide some clues to his or her behavior because ordinal position affects the child's perception of and response to the world.

The *oldest child* has the parent's undivided attention for a period of time, creating a sense of **omnipotence**, or unlimited power or authority. The oldest child may always want things to go his or her way. This perception can lead to difficulties within the family and within the larger community. Parents often have very high expectations for their firstborn. This places great demands on the firstborn.

The *second child* never has the undivided attention of the parents in the same way as the first child. This child has a need to compete with the first child, always wanting to be as good as or better than the older sibling. This may motivate the second child to work harder to achieve. Or the child may give up and settle for less than he or she is capable of attaining. Parents may be more relaxed in their approach to child care.

The *youngest child*, the baby of the family, may gain attention and importance from this position. This can serve as either a positive or a negative influence on his or her development.

The *only child* has only adults for company and role models. How the child handles the presence and attention of adults varies with the individual.

Ordinal position alone cannot be used as a determinant of behavior. The size of the family and spacing of the children may also influence each child in his or her particular position. The gender or sex of the child may influence upbringing. It is unfair to make generalizations regarding the differences or similarities between girls and boys. Each family has its own cultural influences and expectations, which undoubtedly affect a child's perception of gender.

FAMILY PATTERNS

Family patterns can be classified as autocratic, democratic, or laissez-faire, depending on how family members relate to each other.

In the *autocratic family*, parents usually make all decisions. Rules are made and enforced by the adults without input from the children. Parents demand and expect respect from their offspring. The *democratic family* offers its members choices and encourages participation and individual responsibility. This family is based on a philosophy of mutual respect. It is thought that children develop a greater sense of self-esteem and autonomy in the democratic family. The family meeting is an effective tool used by the democratic pattern to air and work out differences. The *laissez-faire family* offers its members complete freedom. Parents do not try to regulate or set limits on the family members. Children raised under the laissez-faire method often do not learn the rules that teach impulse control.

Families may also be considered functional or dysfunctional. A **functional family** is one that fosters the growth and development of its members. Cohesion among family members also helps to promote emotional as well as physical and social well-being. Meeting each family member's needs for love, belonging, and security helps to maintain the stability of the family. The functional family readily admits new members into the circle without compromising the worth and individuality of its members. Healthy families can recognize and accept the differences among individual members and accommodate stresses from inside or outside the family. Common family stresses include financial problems, parenting concerns and conflicts, illness, death, divorce, lack of time, and unequal distribution of roles. Healthy families are not problem-free, but they are able to deal with their problems as a group or seek outside assistance to help them preserve their integrity.

The **dysfunctional family** is unable to offer its members a stable structure. As a result, family members may have poor interpersonal skills and lack the ability to deal with stress and conflict. A lack of proper discipline and consistency can lead to acting-out or antisocial behaviors. Dysfunctional families have trouble reaching outside of the immediate family boundaries for help.

CULTURAL ISSUES

Each family has its own history, structure, and style of functioning. Within the framework of the larger society, groups of families are connected to each other by race, religion, and geographic proximity. Their shared beliefs, values, ideas, and religious doctrines—their common culture—are handed down from generation to generation and adapted or changed to meet the current needs of the group. **Culture** refers to all of the learned patterns of behavior passed down through the generations. During recent years much attention has been given to examining how cultural influences affect health and health care (Fig. 2–4). The nurse can better understand patients' behavior and response to health care by understanding their history and beliefs. Failure to develop cultural awareness may lead to misperceptions about patients' feelings and responses. These misunderstandings can increase the stress for both patient and caregiver. Shared values give a culture stability and security. Culture gives groups of individuals a style of thinking, a way of organizing, and a guide for human interaction. A person's culture determines what values or achievements are important (e.g., independence, work roles, or leisure). Male and female roles are defined by the cultural group to which one belongs. Because culture determines our thinking and behavior, it is an essential force in health care. Culture influences diet, eating practices, how we raise our children, our pain perception, and our reactions to stress and death.

The United States has a broad culture composed of shared values. Within this broad culture many subcultures exist. Members of each subculture retain their fundamental cultural practices and beliefs as seen in the retention of their native language and ethnic celebrations. Today the United States is no longer considered a "melting pot." Today we celebrate our diversity. Our population is very heterogeneous, consisting of many ethnic groups: European Americans, African Americans, Hispanics, Asians and Pacific Islanders, Native Americans, and others. Any generalization or stereotyping regarding these groups of people can be dangerous because exceptions exist within each one.

African Americans make up 12 percent of the total U.S. population. They are the largest minority group. Members of this group live in all areas of the country and are represented in all socioeconomic levels. One-third live in poverty; this represents three times the poverty rate of the white population. Life expectancy for African Americans has been lower than for the total population throughout the century.

The second largest minority group is made up of Hispanic Americans of Mexican, Puerto Rican, Cuban, Dominican, and Central and South American origin.

FIGURE 2–4
Culture and traditions are important to growth and development.

In 1990 they constituted about 8 percent of the total population and are considered the fastest-growing minority group.

The third largest minority group accounts for about 11 million Asians and Pacific Islanders. This group speaks 30 different languages and has as many different cultures and beliefs. Language and cultural and financial difficulties may make access to health care problematic for this group of individuals.

The smallest minority group consists of Native Americans and others. This group too is characterized by diversity. There are over 400 recognized nations and numerous tribes, each with its own cultural beliefs. Unlike other ethnic groups, this group does not die primarily of heart disease and cancer because these are diseases of old age and a large proportion of this population dies before the age of 45. Alcohol-related disorders affect Native Americans more frequently than they do other groups.

Understanding culture helps the practical nurse better understand the differences that individuals bring to health care. Each culture holds different beliefs about health and illness, life and death, young and old, and right and wrong. Respect, tolerance, and an appreciation of diversity are the tools necessary to guide the nurse in clinical practice. The nurse should listen and learn about each patient's culture before intervening. Nurses must remember to analyze their own behavior and be alert for signs of prejudice and bias, taking care to make certain that these do not influence their approach to patient care. See Box 2–2 for an exercise in religious sensitivity.

BOX 2–2

Religious Sensitivity Exercise

The following questions should help to heighten your sensitivity to your own religious beliefs:

1. Are you affiliated with any religious organization?
2. How important is religion to your daily activities?
3. What helps you renew your strength and hope?
4. Is religion a source of comfort to you?
5. What religious practices do you adhere to in your own life?
6. How do you feel about other religions?

SUMMARY

1. A current definition of family is two or more people who have chosen to live together and share their interests, roles, and resources. All families are bound together by attachment and commitment.
2. Each family is unique, but all families share the goals of survival and personal fulfillment of family members.
3. Basic functions of the family are physical maintenance of family members, protection, nurturance, socialization and education, reproduction, and recreation.

4. Families may go through distinct stages of development: the couple stage, the childbearing stage, the grown-child stage, and the older-family stage.
5. Birth order may influence the child's development.
6. Families may be classified as autocratic, democratic, or laissez-faire, depending on how family members relate to each other.
7. Culture refers to learned patterns of behavior. Today's society is culturally diverse; the nurse must be open and nonjudgmental in working with people of different backgrounds.

CRITICAL THINKING

Mavis Citro, 38 years old, was summoned to an interview with the school nurse because Reginald, her 9-year-old son, had had several altercations with his peers and his teachers. During the interview the school nurse discovered that Ms. Citro and her first husband had divorced when Reginald was 5 years old. Two years later she began seeing another man, also divorced and custodial father of two older boys. After a year-long courtship, the couple got married; shortly thereafter they bought a new home and moved out of the neighborhood. Almost immediately, Reginald, then age 8, started having conflicts with his stepfather and stepbrothers. These conflicts have been escalating lately.

1. What is Reginald's current family type called?
2. Give two reasons for the conflict between Reginald and his new stepfamily.
3. What can be done to establish harmony in the family?

Multiple-Choice Questions

1. Which of these descriptions provides a modern definition of a family?
   a. Two or more people who live together and share a bond of love and intimacy
   b. Two or more people who are related by blood, live together, and share the same values
   c. Two or more people who live together and share common bonds
   d. Two or more people who are related by adoption and share the same ethnicity

2. Which characteristic do all families have?
   a. A specific purpose
   b. Specific roles for their members
   c. A specific number of members
   d. Specific behavioral regulations

3. Which goal is common to all families?
   a. Disciplined action
   b. Ritual acts within the group

c. Monetary success
d. Personal fulfillment of the members

4. Which of the following is a basic family function?
   a. Philosophical ideals
   b. Honesty
   c. Protection
   d. Creativity

5. Which is a common family type?
   a. Open
   b. Closed
   c. Bonded
   d. Extended

• **Student Activity:** Family Observation

Select a family that you can observe closely for a brief period of time. While observing the interactions of the family members, try to answer the following questions.

1. What is the specific role of each family member?
2. What are three strengths unique to this family?
3. What are two outside support systems available to this family?
4. What stresses can be identified during this observation?
5. What three interventions might enhance this family's coping abilities?

• **Case Study:** A Preschooler's View of Death

The parents of 4-year-old Laura prepared her for the death of a seriously ill uncle. They spoke about illness and answered all questions honestly. When the uncle died, it seemed appropriate for the whole family to attend the wake. In preparation, the parents explained that Uncle George was going to heaven. Several days after attending the wake, Laura asked, "Do you think Uncle George got where he was going yet?" Questions like this one or others, such as "Will he be cold?" or "How does he breathe?", indicate that the concept of death is too complex for young children to comprehend at the moment. It is important that parents recognize these questions as cues to the child's own concern about his or her safety and place within the family.

Suggested Readings

Anderson, K, and Tomlinson, P: The family health system as an emerging paradigmatic view for nursing. Image 24:57–63, 1992.
Edelman, CL, and Mandle, CL: Health Promotion throughout the Lifespan. Mosby-Year Book, St. Louis, 1990.
Ford-Gilboe, M, and Campbell, J: The mother-headed single-parent family: A feminist critique of the nursing literature. Nurs Outlook 44:173–183, 1996.

Friedman, M: Family Nursing Theory and Practice. Appleton & Lange, Norwalk, CT, 1992.
Grossman, D: Cultural dimensions in home health nursing. Am J Nurs 96(7):33–36, 1996.
Health United States: Healthy People 2000 Review. Department of Health and Human Services Pub. No. (PHS) 93-1232. Hyattville, MD, 1992.
McCool, W, Tuttle, J, and Crowley, A: Overview of contemporary families. Critical Care Nursing Clinics of North America 4(4):549–558, 1992.
Schuster, C, and Ashburn, S: The Process of Human Development: A Holistic Life-Span Approach. JB Lippincott, Philadelphia, 1992.
Walsh, F: Normal Family Processes. Guilford, New York, 1993.
Whaley, L, and Wong, D: Nursing Care of Infants and Children. Mosby-Year Book, St. Louis, 1991.
Whall, A, and Fawcett, J: Family Theory Development in Nursing: State of the Science and Art. FA Davis, Philadelphia, 1991.
Zimmerman, S: Family Policies and Family Well-Being: The Role of Political Culture. Sage, Newbury Park, 1992.

Chapter 3

Chapter Outline

Theories of Growth and Development

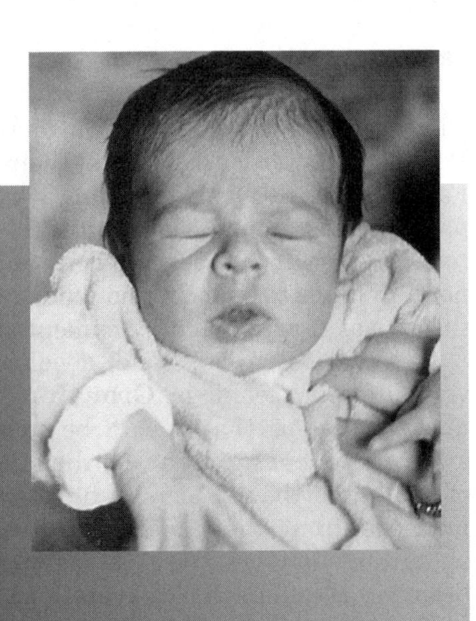

Key Words

autonomy
cephalocaudal
development
ego
ego integrity
Electra complex
generativity
growth
heredity
id
libido
maturation
Oedipus complex
personality
proximodistal
puberty
regression
stagnation
superego

Learning Objectives

At the end of this chapter, you should be able to:

- Describe the five common characteristics of growth and development.
- Name the two major influences on an individual's growth and development.
- Compare Freud's psychoanalytic and Erikson's psychosocial theories of development.
- Describe Jean Piaget's theory of cognitive development.
- Describe Kohlberg's theory of moral development.
- Describe Maslow's theory of human needs.

Growth and development have always been a topic of interest for many individuals. People are always curious about their beginnings and about what their future holds. Families are always questioning why one child in the family looks more like one parent than another. The healthcare worker needs to understand the normal patterns of growth and development and learn to recognize any variations from the norm in order to support and guide parents. Several characteristics, patterns, and theories of growth and development are explored in this chapter.

The terms "growth" and "development" are frequently used together but have very different meanings. **Growth** refers to an increase in physical size. Growth is quantitative in that it can be measured in inches, centimeters, pounds, or kilograms. **Development**, on the other hand, refers to the progressive acquisition of skills and the capacity to function. Development is qualitative in nature and proceeds from the general to the specific. Growth and development occur simultaneously and are interdependent. Development results from learned behavior as well as from maturation. Maturation is similar to development and is a total process that involves the unfolding of skills and potential regardless of practice or training. **Maturation** is the attainment of full development of a particular skill.

Two directional terms used to explain growth and development are "cephalocaudal" and "proximodistal." **Cephalocaudal** is best described as growth and development that begins at the head of the individual and progresses downward toward the feet. **Proximodistal** describes growth and development that progresses from the center of the body toward the extremities (Fig. 3–1). In the infant, shoulder control precedes mastery of the hands, which is followed by finger dexterity.

As we discussed in Chapter 1, health is influenced by both genetic and environmental factors. Genetics and environment are also the major influences on the individual's growth and development. Genetics, or **heredity**, includes all the characteristics such as hair color, eye color, and body size and shape. A more detailed discussion of heredity can be found in Chapter 4 (p. 59).

Certain assumptions are universally accepted as characteristic of growth and development.

1. Growth and development occur in an orderly pattern from simple to complex; one task must be accomplished before the next one is attempted. For example, infants must learn head control before they can learn to sit.
2. Growth and development are continuous processes characterized by spurts of growth and periods of slow, steady growth. For example, infancy is a period of very rapid growth; after infancy, the rate of growth slows down until adolescence.
3. Growth and development progress at highly individualized rates that vary from child to child. Individuals have their own growth timetable, and one child's pattern of growth should not be compared to another's.
4. Growth and development affect all body systems but at different time periods for specific structures. Although many organs mature and develop throughout childhood, the reproductive organs mature at puberty.
5. Growth and development form a total process that affects the person physically, mentally, and socially.

Every person goes through certain stages of development from infancy to old age. As individuals progress through these stages, they are exposed to different environmental factors that influence their inherited makeup. The resultant behavior is

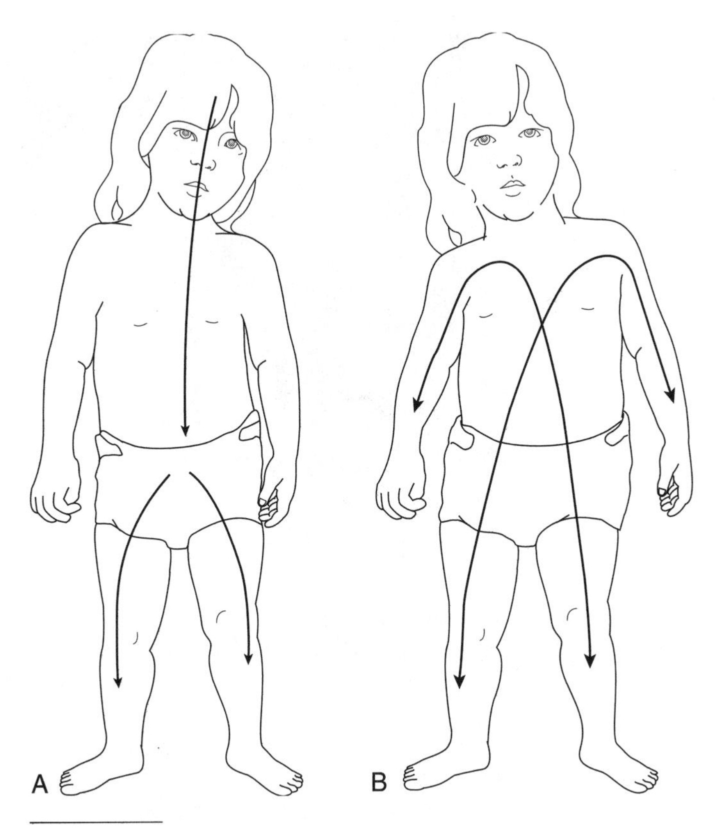

FIGURE 3–1
Principles of growth and development. (*A*) Cephalocaudal. Growth and development proceed from head to toe or tail. (*B*) Proximodistal. Growth and development proceed from the center outward.

unique to that person and is known as personality. **Personality** consists of the behavior patterns that distinguish one person from another—the individual's style of behavior (Fig. 3–2). Personality traits stay identifiable throughout the person's life span. A solid understanding of personality development can assist the health-care worker in the promotion of health and in the delivery of care.

Although no single theory explains the personality development of all individuals, several major theorists provide key frameworks that help nurses understand different aspects of personality development. We include a brief overview here of Sigmund Freud's psychoanalytical theory, Erik Erikson's psychosocial theory, Jean Piaget's cognitive theory, Abraham Maslow's human needs theory, and Lawrence Kohlberg's theory of moral development.

Most of these theories are covered in greater depth in the chapters that follow. Freud's theory provides the foundation from which other theories developed. We chose to present his theory in this chapter because nurses need to have a basic knowledge of personality development. This will enable them to identify the behaviors that are associated with the various stages and better understand whether the behavior is appropriate or inappropriate for a particular developmental level. Unlike later theorists, Freud believed that infancy and childhood are the critical periods for development and change. Freud's theory discusses only these

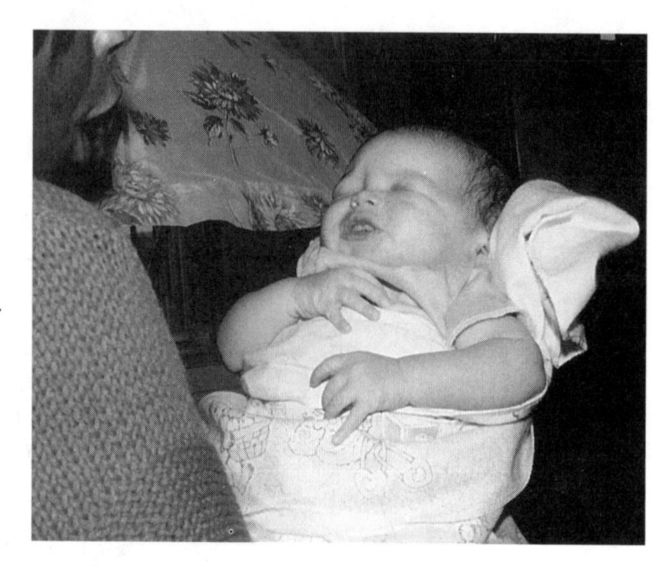

FIGURE 3–2
Each infant has a unique personality.

stages. As authors, we believe that development is ongoing throughout the life cycle, and therefore we limited the discussion of Freud to this chapter alone but refer to other theorists in later chapters because their theories apply to each stage of growth and development.

All developmental theories are divided into stages and are considered progressive. Ideally, a task or skill is accomplished at one stage before moving on to a later stage. However, conflicts and stresses can delay or prolong the completion of a task or even cause some temporary backward movement, known as **regression**. After the resolution of the conflict or stress individuals usually return to their appropriate developmental level. The specific age ranges given for these developmental stages are approximate and vary somewhat for individuals. It is even possible for stages to overlap, allowing individuals to work on several tasks at the same time.

PSYCHOANALYTICAL THEORY

Sigmund Freud made many important contributions to the understanding of personality development. Three components of his theory include levels of awareness, the components of the personality or mind, and psychosexual stages of development. According to Freud, the levels of awareness include the conscious, subconscious, and unconscious. The conscious level refers to all those experiences that are within one's immediate awareness. It is reality-based and logical. The subconscious or preconscious level of awareness stores memories, thoughts, and feelings. These can be recalled with a little effort and brought into the conscious level. The unconscious level refers to that part of the mind that is closed to one's awareness. These stored memories are usually painful and are kept in the unconscious to prevent anxiety and stress. Freud believed that behavior could be understood by delving into the forces of the unconscious mind. The levels of awareness became the basis for Freud's theory of psychoanalysis.

Freud further believed in the three functional components of the mind known as the id, the ego, and the superego. The **id** refers to the body's basic primitive urges. Primarily concerned with satisfaction and pleasure, the pleasure principle, or **libido**, is the driving force behind all kinds of human behavior. The id operates according to the pleasure principle. The id demands immediate satisfaction of its drives. The **ego**, also known as the "executive of the mind," is the part that is most closely related to reality. This part develops as a result of the demands of the id and the forces in the environment. Through interactions with the environment the child learns to delay immediate satisfaction of its needs. This learned behavior is the development of the ego. The **superego** is a further development of one's ego. It makes judgments, controls, and punishes. It dictates right from wrong and acts in a way that is similar to what is thought of as a conscience. These three components, id, ego, and superego, are in constant conflict with one another. Ideally, a balance or compromise should be reached between them. Someone once attempted to explain what each of these components is trying to communicate. The id says, "I want it now!" The superego states, "You can't have it." And the ego attempts to compromise by saying, "Well, maybe later." Unrestrained id dominance can result in a breakdown of the personality, leading to childlike behavior persisting throughout adult life. An extremely harsh superego can cause the blockage of reasonable needs and drives. See Figure 3–3 for Freud's components of the mind.

Freud described five stages of psychosexual development: oral, anal, phallic, latency, and genital. Each stage is associated with particular conflicts that must be resolved before the child can move on to the next stage. He also believed that the experiences a child has during the early stages of growth determine later adjustment patterns and personality traits in adult life (Table 3–1).

Oral Stage

The oral stage lasts from birth to the end of the first year of life. The infant's mouth is the source of all comfort and pleasure (Fig. 3–4). If the infant's oral needs are met, the infant gains satisfaction. The infant receives pleasure by sucking and biting, using the mouth as its center of gratification. By the end of the first year of life, the infant begins to see that he or she is separate from the mother and other objects in the environment.

Anal Stage

The anal stage lasts from the end of the first year of life to the third year. At the beginning of this stage, the mouth continues to be an important source of satisfaction for the child. By the beginning of the second year, the center of pleasure is shared between the mouth and the organs of elimination. Instead of being repulsive to the child, the process of elimination gives the child pleasure and satisfaction. Toilet training is initially experienced as a conflict between the demands of the parent and the child's biological needs. Resolution of this conflict gives the child a sense

ID EGO SUPER EGO

FIGURE 3–3
Freud's three functional components of the mind.

TABLE 3–1
FREUD'S STAGES OF PSYCHOSEXUAL DEVELOPMENT

Age	Stage	Major Developmental Tasks
Birth–18 months	Oral	Relief from anxiety through oral gratification of needs
18 months–3 years	Anal	Learning independence and control, with focus on the excretory function
3–6 years	Phallic	Identification with same-sex parent; development of sexual identity; focus on genital organs
6–12 years	Latency	Sexuality repressed; focus on relationships with same-sex peers
13–20 years	Genital	Libido reawakened as genital organs mature; focus on relationships with members of the opposite sex

From Townsend, MC: Psychiatric Mental Health Nursing: Concepts of Care, ed 2. FA Davis, Philadelphia, 1996, p. 44. Reprinted with permission.

of self-control and independence. Recommendations for toilet training will be discussed in a later chapter.

Phallic Stage

The phallic stage lasts from age 3 to age 6. At this stage, the child associates both pleasurable and conflicting feelings with the genital organs. During this period the child devotes a lot of time to examining his or her genitalia. Masturbation and interest in sexual organs are normal. Exhibitionism is also typical at this age. The child appears quite comfortable with his or her body and likes to undress and parade around naked. Parental disapproval of the child's preoccupation with the genitals can result in feelings of confusion and shame. The Oedipus and Electra

FIGURE 3–4
Infants derive pleassure and comfort from sucking.

complexes develop at this stage. The **Oedipus complex** refers to a boy's unconscious sexual attraction to his mother. He wishes to have his mother to himself and sees his father as a rival for his mother's affection. To win his mother's affection, he resolves the conflict by eventually taking on the father's characteristics. This process begins sex-role identification. The **Electra complex** occurs when a young girl is attracted to her father and wishes to get rid of her mother. Through imitation the child copies the mothering role and eventually gains the father's affection and approval. Resolution of the Electra complex produces sex-role identification for the female child.

Latency Stage

Latency lasts from age 6 to about age 12. During this time the child's sexual urges are dormant. The sexual energies are being channeled into more socially acceptable means of expressions. School-age children focus mainly on intellectual pursuits. Peer relationships intensify between children of the same sex. Sports and other such activities help in the development of these peer relationships (Fig. 3–5).

Genital Stage

The genital stage begins with the onset of **puberty**. During puberty many physical changes occur that prepare the body for reproduction. The hormonal activity and maturing of the sex organs result in the awakening of sexual attraction and interest in heterosexual relationships. The child continues to struggle with a desire for independence but still has a need for parental supervision.

PSYCHOSOCIAL THEORY

Erik Erikson, a psychologist and close follower of Freud, broadened Freud's theory of personality development. Erikson identified eight stages that span the full life cycle from infancy to old age. He studied the child within a larger social setting,

FIGURE 3–5
Sports are important to the school-age child.

beyond the immediate family. He believed that at each stage certain critical tasks have to be accomplished. The successful completion of each task enables individuals to increase independence and feel good about themselves and others. Erikson's eight stages of psychosocial development are discussed below and found in Table 3–2.

Trust versus Mistrust (Birth to 18 Months)

At birth the child is helpless and totally dependent on others to meet his or her needs. When these needs are met in a timely fashion, the child develops trust in people and in his or her environment. Trust is the foundation of the healthy personality.

Autonomy versus Shame and Doubt (18 Months to 3 Years)

The child begins to gain control over his or her body and develop a sense of independence or autonomy (Fig. 3–6). **Autonomy** is characterized by the acquisition of

TABLE 3–2
STAGES OF DEVELOPMENT IN ERIKSON'S PSYCHOSOCIAL THEORY

Age	Stage	Major Developmental Tasks
Infancy (Birth–18 months)	Trust vs. mistrust	To develop a basic trust in the mothering figure and be able to generalize it to others
Early childhood (18 months–3 years)	Autonomy vs. shame and doubt	To gain some self-control and independence within the environment
Late childhood (3–6 years)	Initiative vs. guilt	To develop a sense of purpose and the ability to initiate and direct own activities
School age (6–12 years)	Industry vs. inferiority	To achieve a sense of self-confidence by learning, competing, performing successfully, and receiving recognition from significant others, peers, and acquaintances
Adolescence (12–20 years)	Identity vs. role confusion	To integrate the tasks mastered in the previous stages into a secure sense of self
Young adulthood (20–30 years)	Intimacy vs. isolation	To form an intense, lasting relationship or a commitment to another person, cause, institution, or creative effort
Adulthood (30–65 years)	Generativity vs. stagnation	To achieve the life goals established for oneself, while also considering the welfare of future generations
Old age (65 years–death)	Ego integrity vs. despair	To review one's life and derive meaning from both positive and negative events, while achieving a positive sense of self-worth

From Townsend, MC: Psychiatric Mental Health Nursing: Concepts of Care, ed 2. FA Davis, Philadelphia, 1996, p. 47. Reprinted with permission.

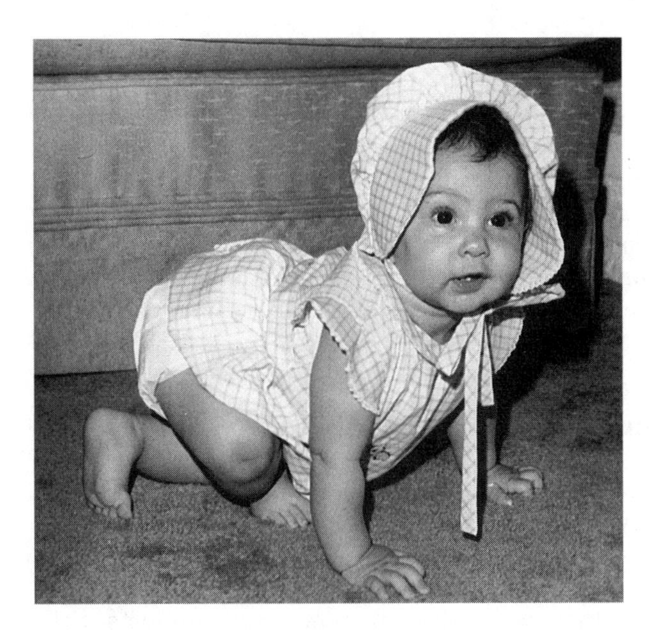

FIGURE 3–6
At 18 months, the infant gains autonomy and some mobility.

skills involving feeding, mobility, dressing, and control of elimination. Developing independence strengthens the child's self-concept. Without loving support from the environment, the child develops feelings of shame and doubt.

Initiative versus Guilt (3 to 6 Years)

During this stage the child begins to explore his or her environment and try different roles. Imagination and curiosity allow the child to further expand and develop his or her potential. Parents and caregivers need to permit the child to explore within safe boundaries. Without this freedom, the child may develop guilt and feelings of inadequacy.

Industry versus Inferiority (6 to 11 Years)

During this stage the child acquires many new social and physical skills. School-age children have the maturity to concentrate on learning and working with others. They strive for praise and recognition. Without these positive responses they may develop a sense of inferiority.

Identity versus Role Confusion (12 to 20 Years)

This stage is transitional between childhood and adulthood. It is characterized by both physiological and emotional changes that create turmoil for both the child and the family. One of the chief concerns of this period is the individual's emerging sexuality and the need to find his or her place in society. Many demands are placed on the adolescent in terms of career, vocation, education, and peer relationships. Role confusion results if the individual does not have love and support.

Intimacy versus Isolation (20 to 30 Years)

A goal of this stage is to establish a close meaningful relationship with another person. The individual must be able to give of himself or herself and be committed to another. This is learned from within the family unit during the growing years. Close ties with family members and intimate relationships are essential to the well-being of the young adult. Failure to accomplish a meaningful close relationship results in loneliness and isolation. Some individuals have many superficial relationships that leave them unfulfilled. Commitment and drive are also needed for career choice and success.

Generativity versus Stagnation (30 to 65 Years)

Erikson defines **generativity** as the process by which the middle-aged person focuses on leadership, productivity, and concern for future generations. Individuals reflect on their accomplishments and become involved with their new family roles. Generativity takes on different forms. Some adults engage in nurturing their children or grandchildren; others become involved in community projects. Still others begin new careers at this stage. Inability to establish generativity results in stagnation. **Stagnation** occurs when a person is unconcerned with the welfare of others and is preoccupied with himself or herself.

Ego Integrity versus Despair (65 Years and Over)

During this period, life experiences are reviewed. **Ego integrity** is achieved if the person reaches a level where he or she is able to accept past choices as the best that could be accomplished at the time. The individual has a sense of dignity from his or her life accomplishments. Ego integrity implies that the individual has resolved the tasks of earlier stages and has little desire to relive his or her life. Dissatisfaction with life review leads to feelings of despair. The person may wish to start over and have another chance. Despair produces feelings of worthlessness and hopelessness.

COGNITIVE THEORY

Jean Piaget's contribution to the field of psychology is cognitive development. He was concerned with how the individual acquires intellect and develops thought processes. Piaget believed that intelligence was an innate ability that further developed as the child adapted to the environment. He believed that the child's cognitive abilities progress through four stages: sensorimotor, preoperational, concrete operational, and formal operational (Table 3–3).

Sensorimotor Stage (Birth to 2 Years)

At birth the infant begins by responding to the environment primarily through reflexes. Gradually, the infant acquires knowledge through exploring the environment and attaches meaning and reconition of things. Through trial-and-error behavior, the child perfects sensory and motor reflex skills (Fig. 3–7). By the

TABLE 3–3
PIAGET'S STAGES OF COGNITIVE DEVELOPMENT

Age	Stage	Major Developmental Tasks
Birth–2 years	Sensorimotor	With increased mobility and awareness, development of a sense of self as separate from the external environment; the concept of object permanence emerges as the ability to form mental images evolves
2–6 years	Preoperational	Learning to express self with language; development of understanding of symbolic gestures; achievement of object permanence
6–12 years	Concrete operations	Learning to apply logic to thinking; development of understanding of reversibility and spatiality; learning to differentiate and classify; increased socialization and application of rules
12–15 years and up	Formal operations	Learning to think and reason in abstract terms; making and testing hypotheses; ability to think logically and reason expand and are refined; cognitive maturity achieved

From Townsend, MC: Psychiatric Mental Health Nursing: Concepts of Care, ed 2. FA Davis, Philadelphia, 1996, p. 51. Reprinted with permission.

completion of this stage, the child is able to see himself as separate from other objects in the environment .

Preoperational Stage (2 to 6 Years)

The child is concerned with the development and mastery of language. This stage is characterized by egocentrism. The child sees himself or herself as the center of the universe and is unable to accept other viewpoints. He or she uses language skills and gestures to meet his or her needs. At this time, objects are singular and one-dimensional to the child. This means that the child can create a mental picture of an object or person.

Concrete Operational Stage (6 to 12 Years)

The increased acquisition of cognition allows the child to think and converse on many topics. The child is beginning to think logically and problem-solve to some degree but is unable to deal with hypothetical or complex abstract situations. The child is less egocentric and more social. Concepts of reversibility and spatiality are developed. Children at this age can understand that water can be in liquid or solid form and can change back and forth. Children at this stage can classify objects using several characteristics. For example, they see a car not simply as a car but as a 1997 Ford Taurus.

Formal Operational Stage (12 to 15 Years)

The individual has the ability to think logically in hypothetical and abstract terms. He or she demonstrates both form and structure in organizing thoughts and is capable of scientific reasoning and problem solving.

FIGURE 3–7
Trial-and-error practice helps the young child learn a new skill.

HUMAN NEEDS THEORY

Abraham Maslow described human behavior as being motivated by needs that are ordered in a hierarchy (Fig. 3–8). At the bottom are basic survival needs (physiological needs, safety, belonging), and at the top are more complex needs (self-esteem, self-actualization). Maslow believed that people must meet their most basic needs before they can move up the hierarchy to the highest level.

Physiological Needs

The most basic needs are physiological and include the need for oxygen, food, water, rest, and elimination. Maslow also included sexual needs, which are important for survival of the species, among the basic needs. When these needs are met, the individual is free to move to the next stage. However, if these needs are not met, the individual will continue to be preoccupied with them. For example, a hungry child may lack interest in school (or anything other than food) until he or she is no longer hungry.

Safety Needs

The need to feel secure, safe, and free from danger is the next need to be met. But one cannot think of safety until physiological needs have been met. The young child

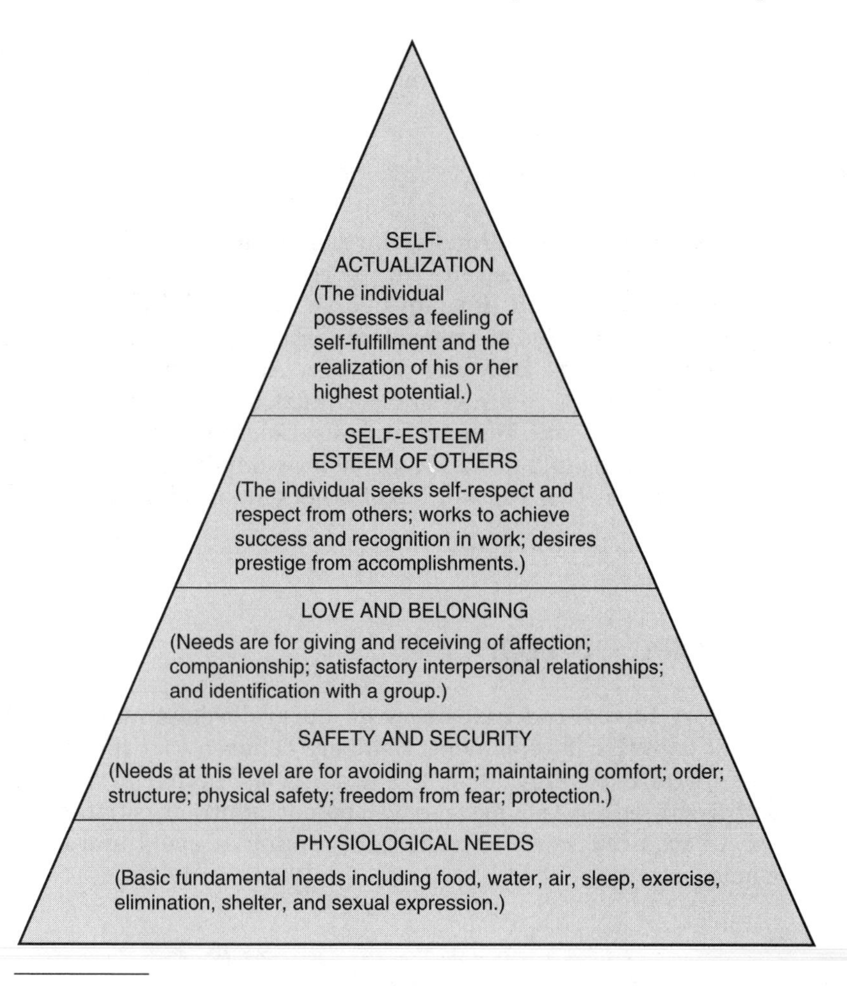

FIGURE 3–8
Maslow's hierarchy of needs (Adapted from Maslow, A: Motivation and Personality. Harper & Row, New York, 1954).

must have feelings of security in the home and family before he or she can venture out into the larger community and school environment.

Belonging

This is the need to feel loved and accepted by another person. In order to enter into any relationship, the person must first feel secure. Love and affection begin with bonding at the time of birth and continue throughout human development. All individuals need affection and meaningful relationships.

Self-Esteem

People need to feel good about themselves and their accomplishments. To arrive at this place, each person must receive approval and recognition of his or her own worth.

Self-esteem is first built by parental approval and acceptance. During the school years teachers and other social contacts can further strengthen a person's self-esteem.

Self-Actualization

Self-actualization means self-fulfillment, the achievement of one's full potential. Maslow does not believe that everyone can be completely self-actualized, but as people continue to achieve and develop healthy relationships, they progress toward this goal. As people move toward self-actualization, they become more comfortable with themselves and who they are. At this level people are self-directed in ideas and actions. Self-actualizers must be reality-oriented, flexible, and able to change as needed. Characteristic of self-actualization is the idea that individuals are part of a group but maintain their own individuality. Creativity, a sense of humor, and respect for the welfare of others are fundamental to this level of achievement.

THEORY OF MORAL DEVELOPMENT

Lawrence Kohlberg introduced his theory of moral development by expanding Piaget's stages of cognitive development. Kohlberg believed that the child progressively develops moral reasoning as he or she gains the ability to think logically. Kohlberg identifies three levels of moral development, which are further subdivided into six stages of acquired moral reasoning, beginning at age 4 and extending to adulthood (Table 3–4).

Level 1: Preconventional Thinking (4 to 10 Years)

The child learns reasoning through the parents' demand for obedience. To avoid punishment, the child begins to recognize right from wrong. A 4-year-old might think, "If I'm mean to my brother, I will be punished and sent to my room."

Level 2: Conventional Thinking (10 to 13 Years)

The school-age child begins to seek approval from society. Kohlberg believed that the child at this stage is influenced by external forces in interactions with his or her peers and environment. A 12-year-old knows that it is wrong to cheat in school and wishes to win the approval of both family and teachers.

Level 3: Postconventional Thinking (Postadolescence)

Adolescents develop their own moral code. Moral reasoning is based on the individual's own principles rather than on external forces. Kohlberg further believed that

TABLE 3–4
KOHLBERG'S STAGES OF MORAL DEVELOPMENT

Level/Age*	Stage	Developmental Focus
I. Preconventional (common from 4 to 10 years)	1. Punishment and obedience orientation	Behavior motivated by fear of punishment
	2. Instrumental relativist orientation	Behavior motivated by egocentrism and concern for self
II. Conventional (common from 10 to 13 years, and into adulthood)	3. Interpersonal concordance orientation	Behavior motivated by expectations of others; strong desire for approval and acceptance
	4. Law and order orientation	Behavior motivated by respect for authority
III. Postconventional (can occur from adolescence on)	5. Social contract legalistic orientation	Behavior motivated by respect for universal laws and moral principles; guided by internal set of values
	6. Universal ethical principle orientation	Behavior motivated by internalized principles of honor, justice, and respect for human dignity; guided by the conscience

From Townsend, MC: Psychiatric Mental Health Nursing: Concepts of Care, ed 2. FA Davis, Philadelphia, 1996, p. 52. Reprinted with permission.

*Ages in Kohlberg's theory are not well defined. The stage of development is determined by the motivation behind the individual's behavior.

some individuals never attain this higher level of moral reasoning. Those who operate at this level usually act according to their internal code of beliefs. Most people stop at a traffic signal even when traffic is clear and they know that no one is watching them.

One of the main criticisms of Kohlberg's work is related to the fact that most of the subjects studied were male; little attention was given to the concerns of women. Later theorists suggested two distinct moral voices, care and justice. Most men are motivated by justice, and most women by caring. Understanding moral development is important for nurses to better understand moral issues that affect patients at different stages in development. An understanding of moral reasoning may also assist nurses in making ethical decisions in clinical practice. See Box 3–1 for a guide to moral decision making.

BOX 3–1

Guide to Moral Decision Making

The following questions will assist you in moral decision making:

1. What characteristics make an act right or wrong?
2. How do rules affect moral acts?
3. What action should be taken in this specific situation?

SUMMARY

1. Growth and development, although often used together, have different meanings. Growth refers to an increase in size; development refers to acquisition of skills.
2. Growth and development occur simultaneously and are interdependent.
3. Maturation is the total process that involves the unfolding of the child's potential, regardless of practice.
4. The two major influences on growth and development are heredity and environment. Hereditary characteristics are all those transmitted by the genes. All other factors that affect the unborn child are environmental.
5. There are five universally recognized basic assumptions about growth and development. They:
   - Progress in an orderly manner from simple to complex
   - Are continuous processes
   - Occur at highly individualized rates
   - Affect all body systems and stages
   - Together form a total process
6. Each individual has a unique behavior known as personality. Different theories of personality development help the nurse promote individuals' health and provide health care.
7. These personality theories describe stages of development. The stages are generally progressive; that is, it is necessary to complete an earlier stage before moving on to the next. However, at times the individual may temporarily regress to an earlier stage.
8. Freud described five stages of psychosexual development: oral, anal, phallic, latency, and genital.
9. Erikson developed a theory of psychosocial development that covers the entire life span. Certain tasks need to be accomplished in each of the eight stages: trust versus mistrust, autonomy versus shame and doubt, initiative versus guilt, industry versus inferiority, identity versus role confusion, intimacy versus isolation, generativity versus stagnation, ego integrity versus despair.
10. Piaget's theory focuses on cognitive development, which proceeds through four stages: sensorimotor, preoperational, concrete operational, and formal operational.
11. Maslow believed that human behavior was motivated by human needs placed on a hierarchy from the most basic to the most complex. These begin with physiological needs and progress to safety, belonging, self-esteem, and self-actualization.
12. Kohlberg's theory of moral reasoning identified three levels of moral development: preconventional, conventional, and postconventional. Moral development progresses within these stages in an orderly sequence. However, one does not attain the highest level of moral reasoning.

CRITICAL THINKING

Jane and Bill bring their 4-month-old baby girl, Tayna, to the hospital. She is admitted for vomiting and dehydration and not permitted anything by mouth. Both parents work and must leave the baby in the nurse's care during the day. Using the information given and the material in this chapter, answer the following questions or problems:

1. Identify Tayna's psychosexual level of development according to Freud's stages.
2. What psychosocial task, according to Erikson, would Tayna be struggling with at this stage?
3. Based on Maslow's human needs theory, what need would be of primary concern to Tayna?
4. List one nursing action that would help Tayna in meeting each of the needs that have been identified.

Multiple-Choice Questions

1. Growth can be defined as:
   a. The progressive acquisition of skills
   b. An increase in cognitive ability
   c. An increase in physical size
   d. The rapid development of language

2. According to Freud, what part of the mind acts as one's conscience?
   a. Id
   b. Ego
   c. Superego
   d. Libido

3. According to Erikson's stages of development, which of the following tasks would Roger, age 9, be completing?
   a. Trust
   b. Industry
   c. Initiative
   d. Autonomy

4. At the completion of Piaget's sensorimotor stage of cognitive development, the child:
   a. Can problem-solve
   b. Can reason hypothetically
   c. Has abstract thinking ability
   d. Recognizes himself or herself as separate

Suggested Readings

Christensen, B, and Kockrow, E: Foundations of Nursing. Mosby, St Louis, 1995.
Dickason, E, Silverman, B, and Schult, M: Maternal-Infant Nursing Care. Mosby, St Louis, 1994.
Erikson, E: The Life Cycle Completed. Norton, New York, 1985.
Gilligan, C: In a Different Voice, ed 2. Harvard University Press, Cambridge, Mass., 1982.
Juneau, PS: Essentials of Maternity Nursing. FA Davis, Philadelphia, 1985.
Kohlberg, L: The Philosophy of Moral Development. Harper & Row, San Francisco, 1981.
Papalia, DE, and Olds, SW: A Child's World: Infancy through Adolescence. McGraw-Hill, New York, 1990.
Townsend, MC: Psychiatric/Mental Health Nursing: Concepts of Care. FA Davis, Philadelphia, 1993.
Whaley, L, and Wong, D: Nursing Care of Infants and Children. Mosby-Year Book, St. Louis, 1991.

Prenatal Period to 1 Year

Chapter 4

Chapter Outline

Prenatal Period to 1 Year

Key Words

acrocyanosis
Apgar score
apnea
blastocyst
bottle-mouth syndrome
cervix
chromosomes
circumcision
cleft palate
colostrum
conception
conscience
deciduous
dental caries
dilation
dominant
effacement
embryo
engrossment
fertilization
fetus
fontanels
genes
involution
karyotype
lanugo
meconium
milia
molding
mongolian spot
morula
neonate
normal physiological weight loss
nystagmus
ova
ovulation
physiologic jaundice
placenta
pseudo-menstruation
recessive
sperm
sutures
teratogens
umbilical cord
vernix caseosa
weaning
zygote

Learning Objectives

At the end of this chapter, you should be able to:

- List three factors that promote a healthy pregnancy.
- Name four factors that may have an adverse affect on pregnancy.
- Describe the steps in prenatal development from fertilization to implantation.
- Describe physical development for infants from 1 to 2 months.
- Describe skin manifestations such as vernix caseosa, lanugo, mongolian spots, milia, and acrocyanosis.
- List five reflexes present at birth.
- Name the normal range for vital signs for the newborn.
- Compare the pattern of fine and gross motor acquisition.
- Give an example of cognitive development for this stage.
- State the process of language acquisition during infancy.
- Describe the nutritional needs of developing infants.
- Describe the advantages and disadvantages of breast-feeding and bottle-feeding.
- Distinguish between the stools of breast-fed and formula-fed infants.
- State the normal sleep patterns for the neonate.
- List three interventions used to promote infant safety.
- Name the immunization schedule for the newborn.
- List two concerns for health promotion during the infancy period.

HEREDITY

Two factors that have a large influence on the health of the developing baby are heredity and environment. Each sperm and ovum contributes 23 **chromosomes** to the new entity, which is a single-celled **zygote**. The sex of the zygote is determined by the combination of X and Y chromosomes. The ovum always contains an X chromosome, while the sperm may contain either an X or a Y chromosome. If the ovum is fertilized by an X chromosome sperm, the zygote will be female; if a Y chromosome sperm fertilizes the ovum, a male zygote will result (Fig. 4–1).

Chromosomes carry the **genes**, which transmit all the genetic information or hereditary characteristics from the parents to the child (Fig. 4–2). The genes are found on strands of deoxyribonucleic acid (DNA) within the nucleus of the cell. Some genes are **dominant**. Dominant genes are capable of expressing their trait over other genes. Other genes are **recessive** and can transmit their trait only if they exist in like pairs. If one gene of a pair is dominant and one is recessive, the dominant gene will exert its influence over the recessive gene. Eye color is an example of a trait that is affected by the dominant-recessive pattern of inheritance. In other words, if a brown eye color gene is paired with a blue eye color gene, the dominant brown gene will govern. Over 700 different diseases are the result of defects carried on recessive genes. Sickle-cell disease, Tay-Sachs disease, and hemophilia are some examples of recessive disorders. In order for a child to inherit a recessive trait or disorder, the child must inherit the recessive gene from both parents. The chromosomal structure of an entity is known as its **karyotype**. Karyotyping—mapping the chromosomal structure—can help predict the transmission of certain genetic disorders and is useful in counseling prospective parents.

ENVIRONMENT

From the moment life begins, the environment begins to exercise its influence on the newly formed entity. Good health practices contribute to the development of a healthy baby. The quality of the mother's diet affects her health and that of her

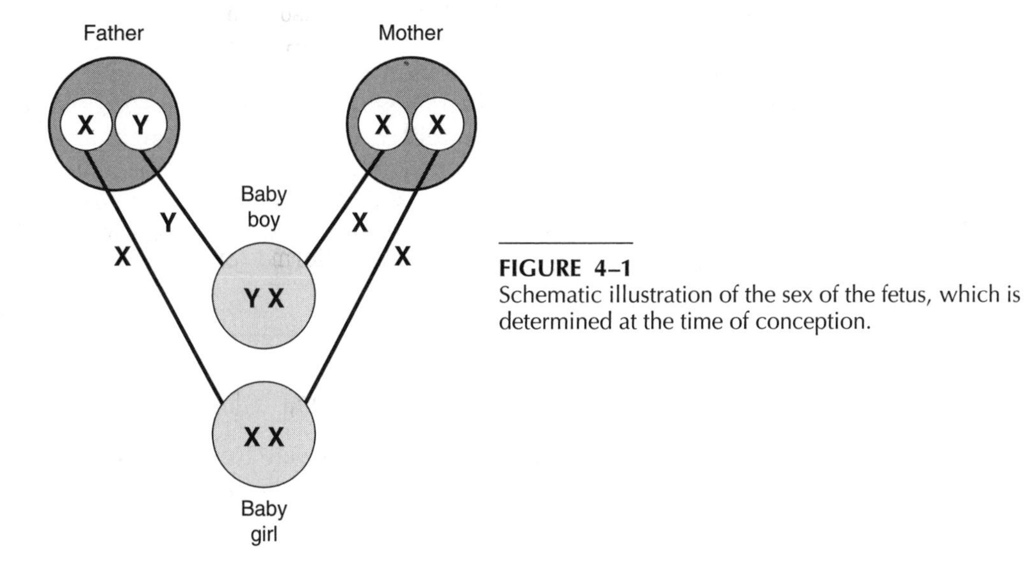

FIGURE 4–1
Schematic illustration of the sex of the fetus, which is determined at the time of conception.

FIGURE 4–2
Hereditary characteristics can be easily seen in a family group.

baby. A balance of rest and exercise is crucial for a healthy pregnancy. As a rule, a woman can continue any exercise that she has regularly participated in before her pregnancy. Walking is the best exercise during pregnancy, and women should be encouraged to walk daily. Before beginning any new form of sport or exercise, the pregnant woman should check with her physician.

Both harmful and life-sustaining substances are transmitted from the mother to the developing baby through the placenta. Chemical or physical substances that can adversely affect the unborn are known as **teratogens**. Tobacco, alcohol, and many drugs are teratogens. As soon as a woman starts to try to become pregnant, she should eliminate all known teratogens to reduce the risks associated with these substances. Bacterial, protozoan, and viral infections may also damage the fetus. The rubella virus presents great risk to the fetus if the woman contracts it during her pregnancy. This virus has been shown to cause serious fetal abnormalities. The human immunodeficiency virus (HIV) may also be transmitted to the unborn child.

Chronic alcohol use may lead to miscarriage or other complications of pregnancy. Infants born to mothers who consume 3 or more ounces of alcohol daily can have fetal alcohol syndrome (FAS). Babies born with fetal alcohol syndrome demonstrate growth retardation, impaired cognition, and congenital malformations. No level of alcohol consumption has been proven safe during pregnancy. Cigarette smoking has also been shown to have teratogenic effects on the unborn. Both low birth weight and growth retardation have been linked to smoking during pregnancy.

The mother should avoid pesticides, chemicals, radiation, and other environmental hazards because of their teratogenic effects. The pregnant woman must follow good health practices and have close medical supervision to ensure her own and her child's health and well-being.

THE PRENATAL PERIOD

The period from fertilization to birth is called the prenatal period. From the time of menarche in puberty until menopause in middle age, the female ovaries produce **ova**, or female sex cells. Roughly every 28 days, an ovum matures and is released in a process known as **ovulation**. From puberty on, the male testes produce **sperm**, or male sex cells, which are released at the moment of ejaculation. Pregnancy begins with the union of the female ovum and the male sperm cell. This is known as **conception** or **fertilization**: all inherited characteristics are determined at this moment.

After fertilization, which normally takes place in the woman's fallopian tube, the zygote undergoes a series of cell divisions and forms a cell mass known as a **morula**. The morula continues to divide and change as it travels down the fallopian tube to the uterus, where it implants itself in the uterine wall. At the point of implantation the entity is called a **blastocyst**. The total process, from fertilization to implantation, takes about 7 days (Fig. 4–3). After implantation, the multicelled structure, now referred to as an **embryo**, continues to develop. By the end of the eighth week of development, all essential structures are formed and the embryo is now termed a **fetus**. The estimated length of pregnancy is approximately 40 weeks (9 calendar months or 10 lunar months). A summary of the process of fetal development is shown in Figure 4–4.

Approximately 280 days after conception (Fig. 4–5), labor begins. Several different hormones are believed to be involved in the process of labor, including progesterone, oxytocin, and prostaglandins. *Progesterone* is produced by the ovaries (female sex glands). It is the hormone that maintains pregnancy and helps stimulate uterine contractions at the end of the pregnancy. *Oxytocin* is produced by the hypophysis, the posterior lobe of the pituitary gland. Oxytocin has two functions: It stimulates uterine contractions and prepares the breasts for breast-feeding. *Prostaglandins* are hormones that are produced in various tissues throughout the body. Like oxytocin, uterine prostaglandins help stimulate contractions.

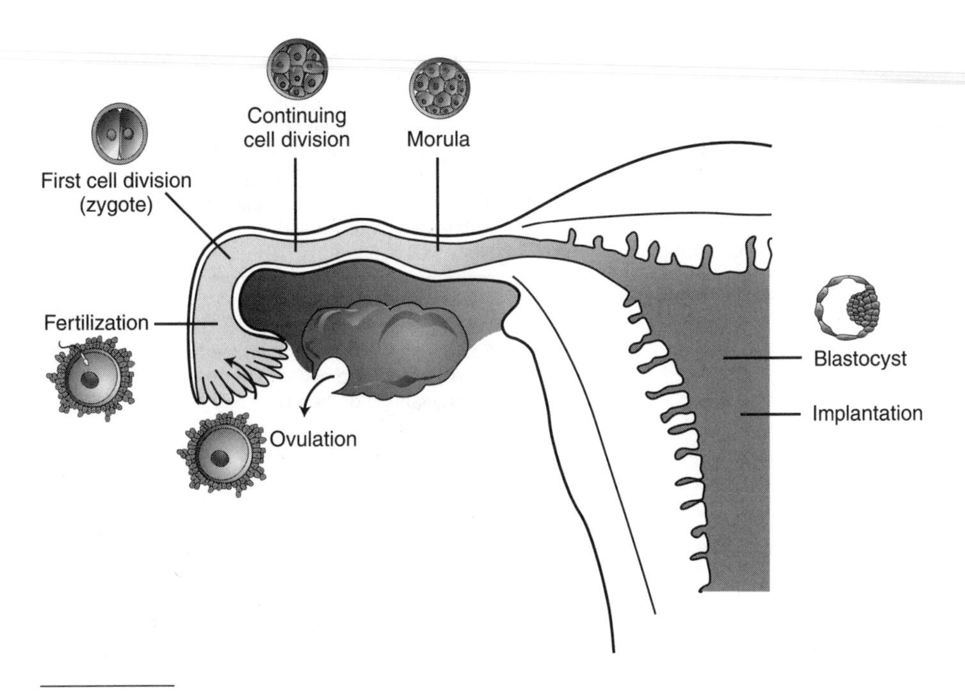

FIGURE 4–3
Ovulation, fertilization, and implantation.

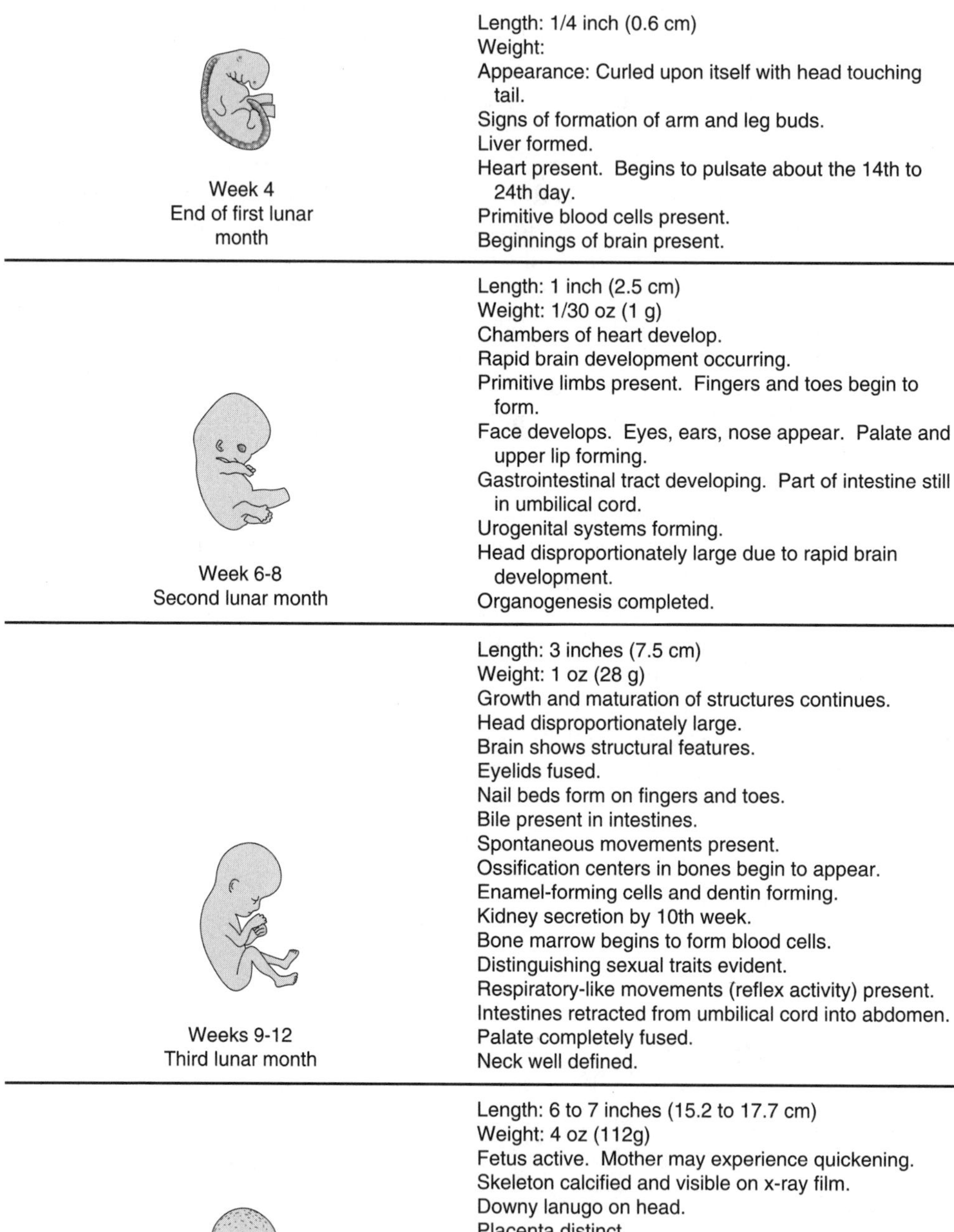

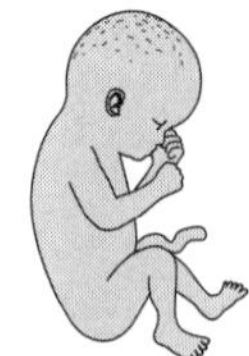

FIGURE 4–4
Fetal development.

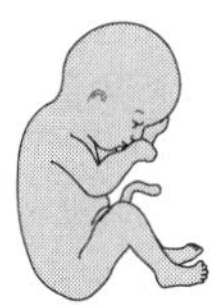

Weeks 17-20
Fifth lunar month

Length: 10 inches (25 cm)
Weight: 8 to 10 oz (224 to 280 g)
Fetal heart sounds evident with stethoscope.
Scalp hair visible.
Lanugo present, especially on shoulders.
Skin less transparent.
Eyebrows present.
Vernix caseosa present.
Fingernails and toenails apparent.
Some fat deposits present.

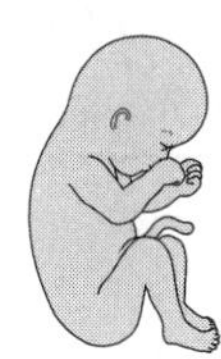

Weeks 21-24
Sixth lunar month

Length 12 inches (30.5 cm)
Weight: 1-1/2 lb (672 g)
Skin wrinkled, pink, translucent.
Increasing amounts of vernix caseosa present.
Eyebrows and eyelashes well defined.
External ear soft, flat, shapeless.
Lanugo covering entire body.
Some breathing effort evident.

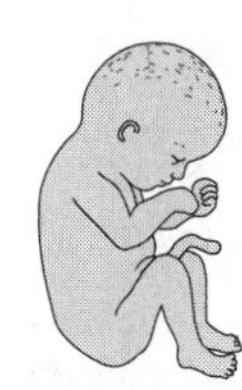

Weeks 25-28
Seventh lunar month

Length: 15 inches (37.5 cm)
Weight: 2-1/2 lb (1120 g)
Skin red, wrinkled, covered with vernix caseosa.
Looks like a "little old man"
Membranes disappear from eyes. Eyelids open.
Scalp hair well developed.
Fingernails and toenails present.
Subcutaneous fat present.
Testes at internal inguinal ring or below.

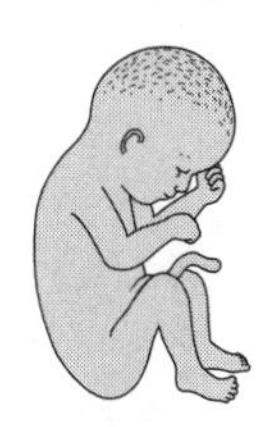

Weeks 29-16
Eigth lunar month

Length: 15 to 17 inches (37.5 to 42.5 cm)
Weight: 3-1/2 to 4 lb (1568 to 1792 g)
Skin pink and smooth.
Areola of breast visible but flat.
Testicles begin descent down inguinal canal (may be in scrotal sac), or
Labia majora small and separated, clitoris prominent.
Hair fine and wooly.
One or two creases evident on anterior portion of soles.
Deposits of subcutaneous fat present.
Can be conditioned to respond to sounds outside of mother's body.

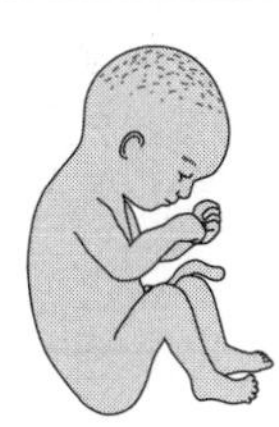

Weeks 33-36
Ninth lunar month

Length: 19 inches (47.5 cm)
Weight: 5 to 6 lb (2240 to 2688 g)
Increased fat deposits give body and limbs a more rounded appearance.
Skin thicker, whiter.
Lanugo disappearing.
Sole creases involve anterior two thirds of foot.
Breast tissue develops beneath nipples.

FIGURE 4–4
(Continued)

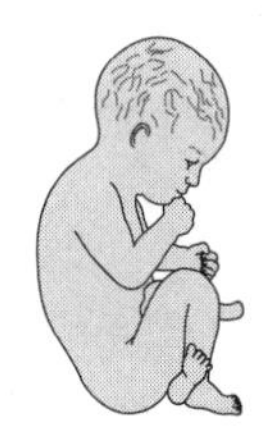

Weeks 37-40
Tenth lunar month

Length: 20 inches (50 cm)
Weight: 7 to 7-1/2 lb (3136 to 3360 g)
After 38 weeks, considered full term.
Body plump.
Lanugo gone from face.
Vernix caseosa disappearing, present in varying amounts.
Testes in scrotum, or
Labia majora meet in midline and cover labia minora and clitoris.
Ear well defined. Erect from head.
Uniform color to eyes (a slate hue).
Acquires antibodies from mother.

FIGURE 4–4
(Continued)

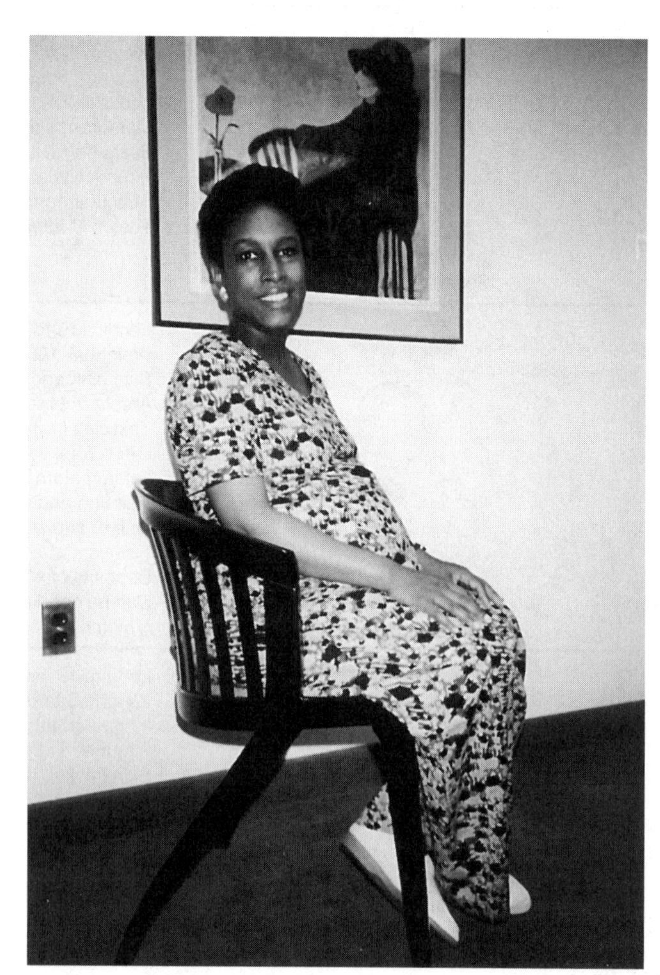

FIGURE 4–5
The approaching birth of her child is an exciting and happy prospect for this mother-to-be. Labor begins 280 days after conception.

There are three distinct stages to labor and delivery. Stage 1, the stage of dilation, is usually the longest, lasting an average of 12 to 24 hours. This stage begins with the onset of regular rhythmic uterine contractions and ends with the complete **dilation** (widening) of the **cervix** (the lower portion of the uterus). During this stage **effacement**, or a shortening and thinning of the cervix, occurs. Stage 2, the expulsion stage, lasts about 1.5 hours, but it is the most difficult stage. It begins with the complete dilation of the cervix and ends with the birth of the baby. Stage 3 is the shortest stage, lasting from 5 to 30 minutes. It begins with the birth of the baby and ends with the delivery of the placenta. The exact length of time for these three stages varies with the individual. Factors such as number of previous pregnancies and deliveries affect the duration of each stage.

Mother and fetus are linked through an organ called the **placenta**. During pregnancy this structure serves many functions, including producing hormones, transporting nutrients and wastes, and protecting the baby from harmful substances. The **umbilical cord** is the connecting link between the fetus and the placenta. At birth it appears to be whitish blue and covered by a glistening membrane. In the delivery room the cord must be assessed for the presence of three vessels—two arteries and one vein. Any deviation from this usually indicates some serious cardiac abnormality. In fetal circulation, oxygen and nutrients reach the fetus by way of the umbilical vein, and waste and deoxygenated blood returns to the placenta for oxygenation by way of the umbilical arteries.

Fetal circulation ends at birth, when the umbilical cord is tied off and the newborn infant, or **neonate**, take its first breath. At 1 minute after birth, and again 5 minutes later, the neonate is assessed on the **Apgar scale** (Table 4–1). Five essential categories of functioning are assessed, including color, reflex irritability, heart rate, respiratory rate, and muscle tone. The Apgar score gives an immediate clinical picture of the newborn's overall status.

TABLE 4–1
APGAR SCORING CHART

Sign	0	1	2	1 Minute	5 Minutes
Heart rate (beats per minute)	Absent	<100	>100	—	—
Respiratory rate (breaths per minute)	Absent	Slow, irregular	Good, crying	—	—
Muscle tone	Limp	Some flexion of extremities	Active motion	—	—
Reflex irritability:					
Response to catheter in nostril	No response	Grimace	Cough or sneeze	—	—
Slap to sole of foot	No response	Grimace	Cry and withdrawal of foot	—	—
Color	Blue, pale	Body pink; extremities blue	Completely pink	—	—

PHYSICAL CHARACTERISTICS

Head and Skull

At birth the newborn's head is large in proportion to the rest of its body, typically one quarter of the total body length. Its average circumference is 13 to 14 inches (33 to 35.5 cm)—about 1 inch larger than the chest. The skull consists of six soft bones: one occipital, one frontal, two parietal, and two temporal bones (Fig. 4–6). The skull bones are separated by bands of cartilage, called **sutures**. Located at the anterior and posterior on the infant's skull are two spaces or soft spots, called **fontanels**. These fontanels are very visible and even appear to pulsate when the infant cries. The skull should be palpated for the presence of sutures and fontanels. The small triangular-shaped posterior fontanel closes by the infant's fourth month. The larger diamond-shaped anterior fontanel closes when the child is 12 to 18 months old. These spaces allow the skull to accommodate the rapid brain growth that takes place in this period. The newborn's skull may appear misshapen or elongated as a result of **molding**, which occurs as the head passes through the narrow birth canal. This is a temporary condition that disappears naturally in a few days.

Length and Weight

The average length of the newborn measured from head to heel is 20 in (50 cm). Normal length for newborns ranges from 19 to 21 in (48 to 53 cm). Usually infants grow 1 in per month for the first year. At 12 months the child's brain is approximately 2.5 times as big as it was at birth, the head and chest are equal in circumference, and the child is 1.5 times longer than at birth.

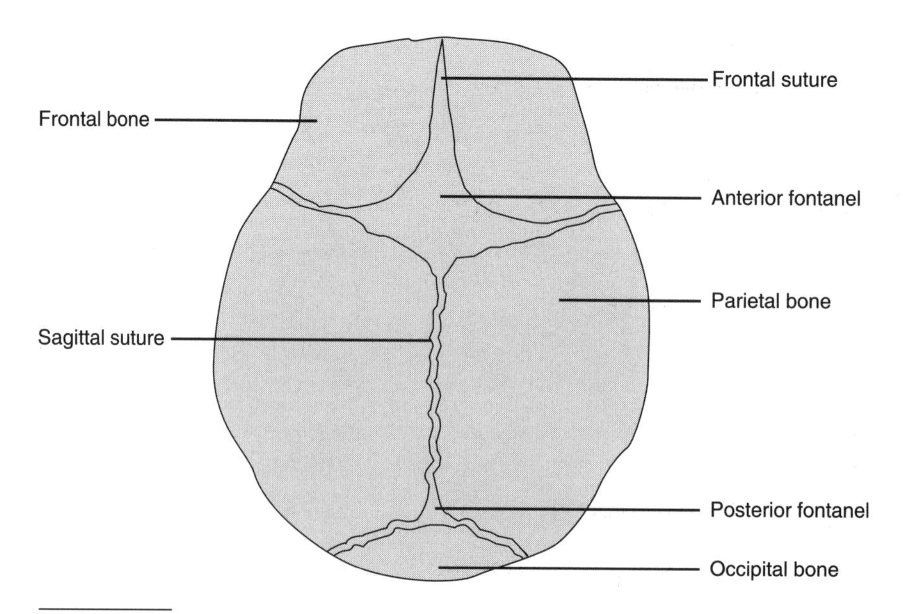

FIGURE 4–6
Infant's skull showing sutures and fontanels.

The newborn's head appears to rest on its chest because its neck is short and has deep creases (Fig. 4–7). The arms and legs are proportionately short and kept in a tightly flexed position. At birth, the newborn weighs an average of 7.5 lb (3400 gm); the normal range is from 5.5 to 10 lb. Boys are generally slightly larger than girls at birth. The newborn loses 5 to 10 percent of its birth weight in the first few days of life. This occurs because the infant is given nothing by mouth for the first few hours and therefore the infant's output exceeds its intake. This is known as **normal physiological weight loss**. When the mother's breast milk comes in or formula feeding begins, the neonate regains its initial weight loss in approximately 10 days. Thereafter the newborn will gain 5 to 6 oz per week for the first month; it will double its birth weight by 5 to 6 months of age and triple its birth weight by its first birthday. Approximately 75 percent of the infant's body weight consists of water. For this reason infants with vomiting and diarrhea may suffer a rapid loss of total body fluid and possible dehydration.

Skin

The newborn's skin at birth is thin and appears pale. Some temporary **acrocyanosis** (blueness of the hands and feet) may be present as a result of poor peripheral circulation; this is usually transient and disappears a few hours after birth. Pigmentation may be more pronounced in certain areas of the body, such as the earlobes, scrotum, and back of the neck; full pigmentation develops several days later. The neonate's color varies according to the amount of melanin present in the skin. In general, infants of northern European descent vary from pink to red; infants of African descent vary from pink to dark red; infants of Asian descent vary from rosy red to yellowish tan; Hispanic- and Mediterranean-descent infants may be yellowish brown; and Native American infants vary from light pink to reddish brown. Infants with relatively more melanin in their skin may be born with a **mongolian spot**, a flat, irregular, pigmented area in the lumbar-sacral region. The mongolian spot usually fades and becomes less noticeable at about age 4.

Many newborns have a covering of fine hair over the body. This covering, known as **lanugo**, vanishes in the first few days after birth. The newborn's skin

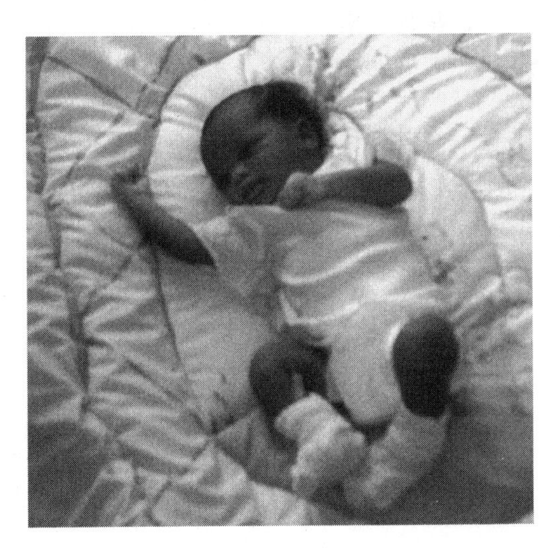

FIGURE 4–7
Physical characteristics of the newborn include short neck and tightly flexed limbs.

creases have a white cheeselike oily covering called **vernix caseosa**, which protects the fetus's skin during pregnancy.

Milia, small clusters of pearly white spots mostly on the infant's nose, chin, and forehead, may also be present at birth. These spots are caused by the retention of sebaceous material within the sebaceous glands and disappear spontaneously without treatment.

Some infants develop a yellow tinge to their skin known as icterus neonatorum, or **physiologic jaundice**. Physiologic jaundice frequently occurs in newborns within 48 to 72 hours after birth. At birth the neonate's red blood cell (RBC) count is higher than that of the normal adult—usually 6 million per cubic millimeter of blood. A few days after birth the RBC count begins to decrease as the body destroys these unnecessary excess cells. This releases a high amount of bilirubin (a component of RBC), producing the jaundiced appearance in the infant. Physiologic jaundice should not be confused with jaundice related to blood incompatibilities, which is usually present immediately at birth and requires prompt medical intervention and treatment.

Genitals

Maternal hormones of pregnancy present in the neonate's bloodstream may cause certain physiologic anomalies. The breasts of neonates of both sexes may be swollen. This condition will disappear without intervention.

The scrotum of male neonates may appear large and edematous. The scrotum should be palpated for the presence of testicles, which usually descend from the abdominal cavity into the scrotal sac during the seventh month of fetal life. If the testicles have not descended, the infant will be observed to see if descent occurs over the next few months. Undescended testicles can be treated with a short course of drug therapy or surgery. The newborn's penis is inspected for the location of the urethral opening. Normally this opening is just at the tip of the head of the penis under the foreskin. **Circumcision**, the surgical removal of the foreskin, may be performed after birth for hygienic or religious reasons. Any deviation should be noted and reported to the physician for follow-up treatment.

The labia in the newborn female may appear swollen. A blood-tinged mucous vaginal discharge known as **pseudomenstruation** may be noted. These conditions are related to maternal hormones and disappear without treatment.

Urine is normally present in the bladder at the time of birth. The newborn should void within 24 hours after birth and 8 to 10 times a day thereafter. The initial voiding may appear rust-colored because of the presence of uric acid crystals. This condition generally disappears without treatment.

Face

The newborn's face is small and the eyes may appear swollen. The eyes are treated after birth with antibiotic application of erythromycin or silver nitrate as a preventive against blindness caused by gonorrhea. Eye color varies from slate gray to dark blue. Permanent eye color is not determined until 3 to 6 months of age. No tears are produced until 4 weeks of age, when the lacrimal ducts (tear ducts) are developed.

The neonate usually has a flat nose and a receding chin. The neonate's mouth is usually examined closely for any defects or abnormalities, particularly **cleft palate**, the incomplete formation and nonunion of the hard palate. This condition can be corrected through surgical repair. The gums should be pink and moist. The first teeth, called **deciduous** or primary teeth, begin to erupt when the infant is about 6 or 7 months old (Fig. 4–8). Usually the first teeth to appear are the two lower central incisors; they are followed by the two upper central incisors. By age 12 months the baby will have between six and eight teeth.

Teething is marked by signs of irritability, excessive drooling, loose stools, and anorexia. The infant can be given hard objects to bite, such as toast, crackers, or zwieback. When the teeth erupt, caregivers should cleanse the new teeth with a soft toothbrush and water. Toothpaste is not recommended because the baby may swallow it. Fluoride supplements are recommended in areas where fluoride is not added to the water supply.

Helpful Hints

Do not use liquor or apply aspirin to the teething infant's irritated and swollen gums.

Abdomen

The neonate's abdomen appears large and flabby. Immediately after birth, the umbilical cord is tied and cut. After a few days the blood vessels of the cord become dry or thrombosed. This is accompanied by a change in color from dull yellowish brown to black. By the tenth day the dried cord falls off and the navel is completely healed. Tub bathing is avoided until the navel is fully healed.

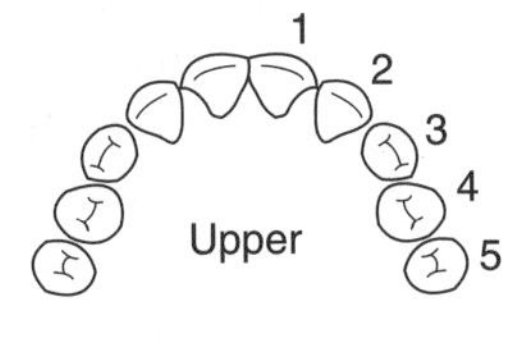

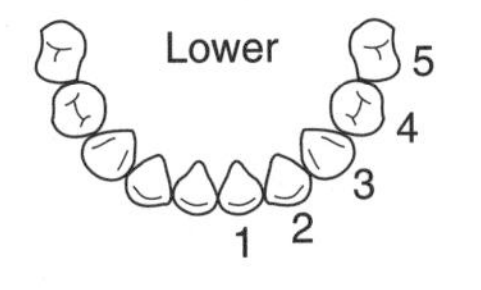

	(Upper)		(Lower)
1 Central incisor	8-12 mos.		5-9 mos.
2 Lateral incisor	8-12 mos.		12-18 mos.
3 Cuspid		18-24 mos.	
4 First molar		12-18 mos.	
5 Second molar		24-30 mos.	

FIGURE 4–8
Approximate ages for the eruption of deciduous teeth.

Helpful Hints

> To hasten cord healing, fold diapers away from the cord stump, apply alcohol to the area around the cord, and report any signs of redness or drainage.

At the time of birth the newborn can swallow, digest, metabolize, and absorb nutrients. The newborn can metabolize only simple carbohydrates, and for this reason whole milk, which contains complex sugars, is not given to the newborn. The newborn's stomach can hold 1 to 3 ounces of fluid; by 10 months, it can hold about 10 ounces. The neonate's cardiac sphincter is underdeveloped, and for this reason, it is important to allow the infant short periods of feeding, followed by "bubbling" or burping for the release of swallowed air.

Bowel movements of healthy infants vary in number, color, consistency, and general appearance. The mother's diet and the type of formula will also influence the infant's stools (Table 4–2). Within 10 hours after birth the newborn should pass its first stool, known as **meconium**. Meconium is thick, green-black, tarry, and odorless. Breast-fed infants have stools that resemble light seeded mustard. Stools of formula-fed infants are commonly semisolid tan or yellowish in color. Some infants have four to six bowel movements a day. An infant is constipated if stools are very hard and can be passed only with much effort. Adding some additional water or strained fruits to the diet may prevent constipation.

Extremities

The newborn's extremities are short in proportion to the rest of the body and kept in a tightly flexed position. They should be examined for range of motion, symmetry, and reflexes. The lower extremities are examined closely to determine if there is an extra gluteal fold, which usually indicates a congenital hip dysplasia (Fig. 4–9). Any abnormality should be reported immediately to the physician for further evaluation. The toes and fingers are counted and inspected for abnormalities. In full-term infants the soles of the feet and palms of the hands are deeply creased. Preterm infants have only very fine lines on their palms and soles.

TABLE 4–2
STOOL PATTERNS FOR NEWBORNS

Stool	Age	Description
Meconium	First 2 days	Greenish-black, tarry, odorless
Transitional	2 to 3 days	Brown to yellow to green
Formula-fed	From day 2 or 3	Pasty yellow or tan, distinct odor
Breast-fed	From day 2 or 3	Light seeded-mustard, sweet odor

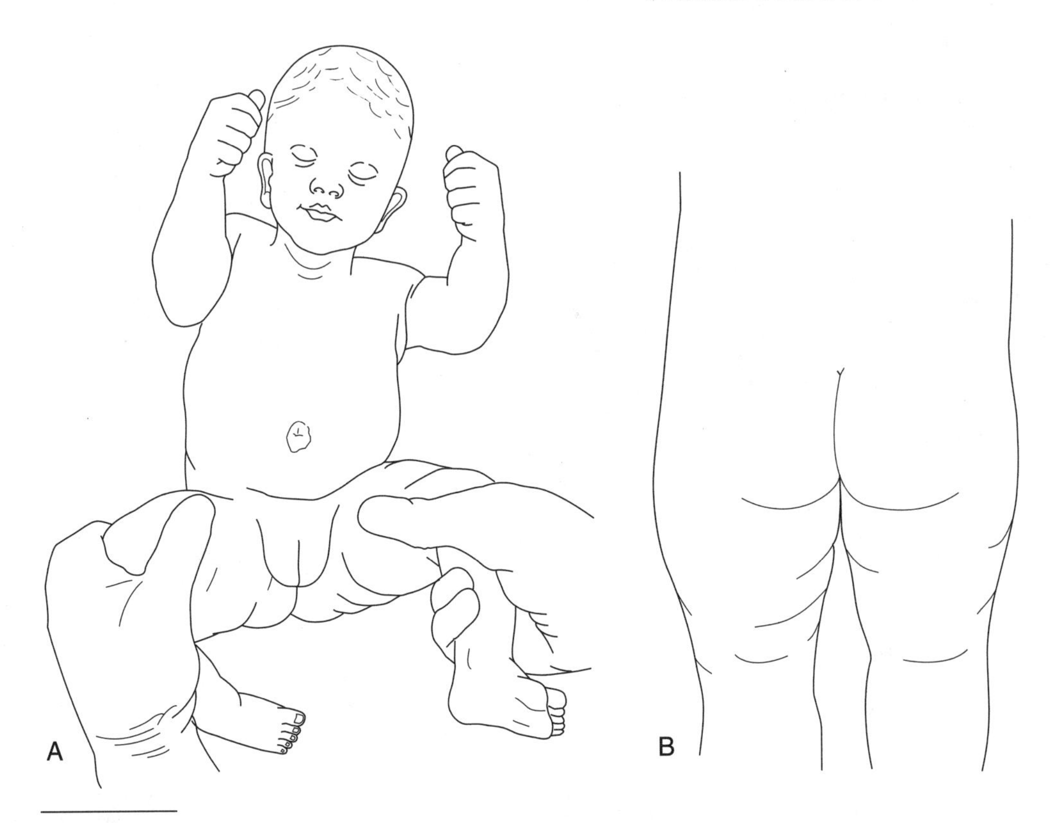

FIGURE 4–9
Assessment of the gluteal and popliteal folds of the hips. The folds should be symmetrical. *(A)* Limitation of abduction. *(B)* Asymmetry of skin folds.

Neurological Characteristics

A neurological assessment in the newborn focuses on reflexes, posture, movement, and muscle tone. At birth, the nervous system is immature and the newborn responds to its environment through a series of reflexes. The presence of certain reflexes indicates a normal neurological system and also helps evaluate approximate gestational age. Several reflexes are protective; these include blinking, sneezing, swallowing, and the gag reflex. Other reflexes present include Moro, or startle reflex, rooting, grasp, Babinski, and tonic neck reflex. Rooting and sucking help the infant secure food. Table 4–3 describes these reflexes in detail.

The newborn's spinal column is inspected to make certain that there are no masses, cysts, or openings. The presence of any spinal defect necessitates immediate medical intervention.

The five senses (sight, hearing, taste, touch, and smell) are present at birth and function at a primitive level. Neonates can track objects at birth; they appear to prefer bright lights and yellow, green, and pink objects, as well as large geometric shapes. The neonate's pupils react to light by dilating and contracting. The newborn's vision is 10 to 30 times less acute than normal adult vision of 20/20. By the time the infant is 6 months old, vision should be 20/100 or better. Movement of the eyes is usually unequal owing to immature cilliary muscles; it is not uncommon for a neonate's eyes to cross or for one eye to drift when focusing on an

TABLE 4–3
REFLEXES PRESENT IN THE NORMAL NEONATE

Reflex	Action	Disappearance/Extinction
Moro	Sudden movement or jarring of position causes extension and adduction of extremities	By age 3 or 4 months
Tonic neck	If head of backlying newborn is turned to one side, infant will extend arm and leg on that side; opposite arm and leg remain tightly flexed	By age 5 months
Rooting	When newborn's cheek is gently stroked, infant turns toward that side and opens mouth	By age 4 to 6 months
Sucking	Newborn makes sucking movements when anything touches lips or tongue	Diminishes by age 6 months
Babinski	When newborn's sole is stroked, toes hyperextend and fan outward; big toe turns upward	By age 3 months
Palmar grasp	Newborn briefly grasps any object placed in hands	By age 3 months (present from age 6 weeks)

object. This is known as **nystagmus**. These deviations are temporary and should disappear without treatment. By 4 months of age infants have binocular vision: they can focus both eyes simultaneously to produce one image. Depth perception at first is limited to grasping for items out of reach and becomes more precise between the ages of 7 and 9 months. At this point the infant is able to reach for items more accurately and purposefully.

The ears are positioned on the sides of the head, with the top of the ear about the level of the eyes. At birth the newborn's ears are generally filled with either vernix or birth fluid, which dissolves within a few days. The infant hears and responds to loud low-frequency sounds. A sudden loud sound will produce a startle response. By the age of 6 to 8 weeks, infants recognize their mother's voice and turn their heads in response to it. A 1-year-old can discriminate between different sounds and often recognize the source.

Newborn infants can discriminate between different tastes. If they are given a sweet solution, they will begin to make sucking movements. When given something sour, they will respond with a grimace or pout. The 1-year-old child has developed a capacity to taste and have preferences for certain flavors. The sweet taste appears to be universally pleasing. The young child should be introduced to a wide variety of tastes and textures. This exposure helps to mature the child's sense of taste.

The sense of touch is keenly present at birth. The face is most sensitive, especially around the mouth, hands, and soles of the feet. Infants like to be touched and rocked because this has a calming effect. Pain perception is present in the newborn and is witnessed when an injection is given. The typical reaction to pain is loud crying and thrusting of the whole body and extremities. The 1-year-old child demonstrates withdrawal from pain but may not be able to recognize the source of the pain. For example, the child may touch a hot pot and quickly respond by withdrawing the hand and crying. However, the child might repeat the action another time, not understanding the cause and effect.

Studies indicate that newborns have a sense of smell. Newborns have been tested and found to react to strong odors by turning away. It has also been documented that newborns can recognize the smell of breast milk. One study showed that infants can even distinguish their own mother's breast milk from others.

VITAL SIGNS

The newborn's temperature immediately after birth may be a little below normal. This is a result of an immature temperature-regulating mechanism and heat loss caused by the cooler environment in the delivery room. The newborn should be dried off and placed under a radiant warmer to help raise its body temperature. In addition, the neonate's head should be covered to prevent further heat loss from evaporation. Once stabilized, the neonate's normal axillary temperature ranges from 97.7 to 99.5°F (36.5 to 37.5°C). The newborn's temperature should be measured using the axillary route to prevent possible rectal perforation.

Pulse should be taken by listening to the chest for an apical pulse for 1 full minute. After stabilization, the apical heart rate ranges from 120 to 140 beats per minute. Slight variations in the heart rate are common. During periods of rest the rate may slow down to 100 beats per minute; during crying periods the rate may increase to 180 beats per minute.

Blood pressure (BP) readings provide a baseline and can be used to assess the infant for cardiac abnormalities. Average BP using oscillometry (Dinamap) is 65/40 mmHg. Blood pressure will increase and heart and respiratory rate will decrease as the child gets older.

Respirations should be counted for 1 full minute. The respirations of the newborn are normally irregular, shallow, and diaphragmatic, with brief periods of **apnea** (absence of breathing). Infant respirations can be counted by watching the abdomen rise and fall. The normal respiratory rate is 30 to 60 breaths per minute. Marked deviations in these normal ranges may indicate congenital abnormalities and will warrant further investigation.

DEVELOPMENTAL MILESTONES

Motor Development

The neonate's movements and behavior appear purposeless and uncoordinated, but all newborns have distinct behavioral characteristics and physical traits that make them different from other neonates. Every infant has his or her own growth timetable. Growth and development should be assessed based on the infant's own individual progress.

Gross Motor Skills

Gross motor skills involve the large muscles of the arms and legs. Following a cephalocaudal pattern, head control develops by 2 months; by 3 months of age the infant can briefly hold the head up. At 4 months the infant can raise the head to a 90-degree angle from the prone position (Fig. 4–10). Rolling over from abdomen to back occurs at 4 months. By 6 months old, the baby can roll both ways, sit with support, and hold the head erect. Sitting alone occurs at the seventh month. The 10-month-old can change position from the prone (face-down) position to the sitting position. Crawling, a primitive movement in which the infant's abdomen is on the floor, is seen in infants at about 9 months of age. Creeping is a more advanced form of movement that requires the infant to raise up on all four limbs (Fig. 4–11). Some infants progress to this style of locomotion by 10 to 11 months. At about 8 months babies can pull themselves up to a standing position. Standing is followed by cruising, which is a form of stepping while holding on to some object or surface for support (Fig. 4–12). Walking unassisted is achieved between the ages of 12 to 15 months. Figure 4–13 outlines developmental milestones.

Fine Motor Skills

Fine motor skills include the fine movements of the hands and fingers. Initially grasp is in the form of a reflex action involving the whole arm in a swiping movement. The neonate exhibits the palmar grasp reflex, grasping any object that is placed in the hands. Finger and hand control develops after shoulder and arm control, demonstrating the principle of proximodistal development. Purposeful reaching and grasping using the whole hand occurs by the fifth month of life. It is common to see infants take hold of objects and immediately bring them to their mouths. For this reason, it is important that safety measures be taken to prevent accidental aspiration of small objects.

The 6-month-old infant is able to hold a bottle, a cracker, or dry toast and bring it to the mouth. At this stage of development, many babies enjoy biting on a hard food object; this can be soothing to their swollen gums. Hand preference usually does not appear until the seventh or eighth month, when the baby can transfer an object from one hand to another. The 7-month-old infant continues to make progress with self-feeding (Fig. 4–14). He or she can hold the bottle and now has the use of the pincer grasp, which permits the opposition of thumb and forefinger. At this time it is typical for babies to grasp and then release small objects. This action usually

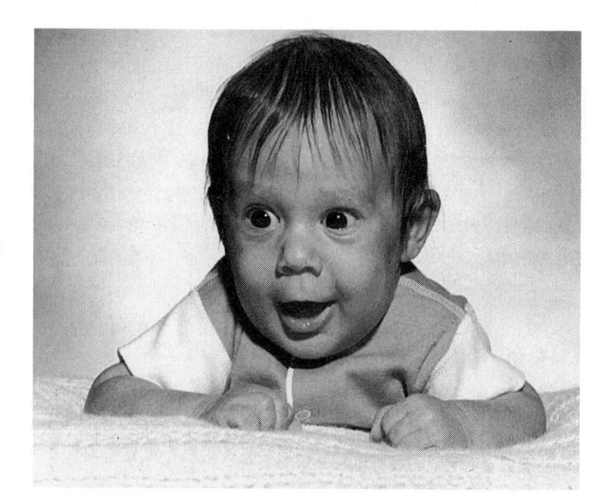

FIGURE 4–10
At 4 months the infant can raise the head to a 90-degree angle from the prone position.

FIGURE 4–11
Creeping is an advanced form of locomotion with the body raised up on all four limbs.

delights the child because it causes the caregiver to retrieve the object so the action can be repeated. By 9 months babies are able to drink from a cup and attempt to use a spoon. In early attempts, the spoon may be inverted and the contents spilled. The 1-year-old can hold a writing object, make scribbling marks on paper or other surfaces, and build a tower of two blocks.

Psychosocial Development

Erik Erikson believed that each child needs to accomplish a particular task at each stage of development. The resolution of each task permits the child to move on to a

FIGURE 4–12
Cruising is an early form of walking while holding on to an object for support.

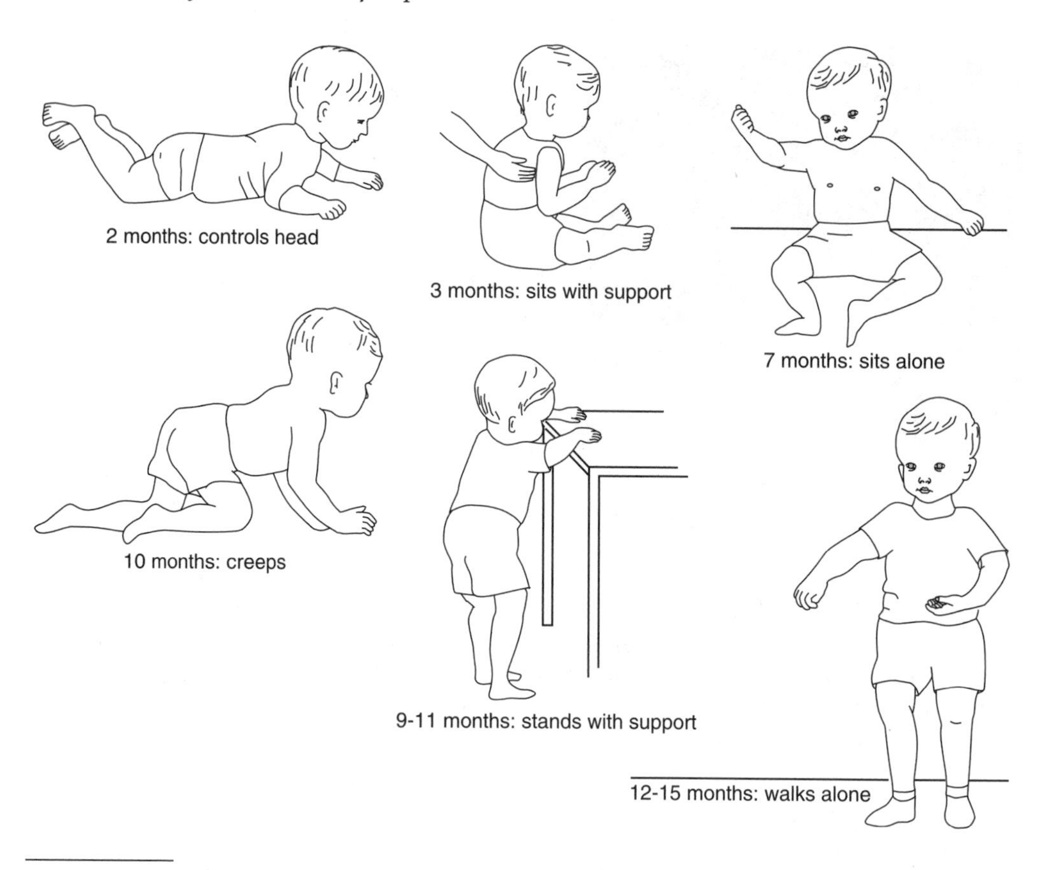

FIGURE 4–13
Developmental milestones.

new stage. For an overview of Erikson's stages of growth and development, refer to Chapter 3. According to Erikson, the infant is working on completing the task of trust. The infant will feel secure and develop a sense of trust when the environment consistently satisfies his or her basic needs for food, comfort, and love. This first stage lays the foundation upon which future stages will be built. Depriving the

FIGURE 4–14
Between 7 and 8 months, infants like to feed themselves.

infant of basic needs can result in the development of mistrust and hinder the further development of the infant's full potential.

The parent-child relationship begins with fetal development and continues after birth. Emotional bonds between the mother and child are known as attachment. This can be evidenced by the way the mother holds, talks to, and looks at the baby. This attachment, or bonding, process serves to strengthen the infant's sense of security and self. The process of bonding, which is also important to the father, is referred to as **engrossment**. Encouraging fathers to participate in the pregnancy and birthing process can initiate the foundation for engrossment. After birth, bonding can be strengthened by involving the father in child care.

Temperament

Babies are born with their own unique temperament, which determines the moods of infants as well as their response to stimulation. Temperament is inborn, whereas personality is shaped and affected by the environment. An infant's willingness to interact with others is a part of its temperament. Some babies are very social and others are shy. Babies with difficult temperaments are more fretful and cry a lot. These infants are not easily soothed. Other infants have an even temperament, which allows them to adapt to their surroundings with little fussing.

Parental Guidance

As infants begin to develop their means of locomotion, the need for discipline increases. For the first 6 months parents may use the art of distraction. The baby who continues to look and reach for the knobs on the stove can be given an age-appropriate toy as a substitute. In the second 6 months, as the infant's memory and cognition increase, discipline must be more direct. The 10-month-old can be told "No" firmly when he or she reaches for something unsafe. The child at this stage is able to understand the tone of repeated admonitions. Verbal cues alone without supervision cannot prevent accidents. Parents and caregivers must be advised that discipline should not be harsh and should focus on praising the positive, desirable behavior while deemphasizing the negative, undesirable behavior. An important goal of discipline is to teach the child impulse control and set limits. The need for discipline will expand and continue throughout childhood and into adolescence.

Cognitive Development

Piaget proposed that the infant begins life with no understanding of the world. The child then must learn about the environment through observation and sensory perception. For example, the baby begins to understand objects by touching, tasting, seeing, hearing, and smelling. Piaget described infancy as the stage of sensorimotor development. Initially, infants respond to stimuli in the environment by reflex action. At about 8 months of age infants begin to plan and coordinate their actions. For example, the infant knows that if he or she shakes a toy, it will produce a sound. By the end of the first year, infants are able to form bonds with certain people and recognize and attach meaning to objects. They begin to be able to understand some repeated actions. For example, the 10-month-old infant has learned that when mother goes to the pantry or refrigerator, she might be getting something to eat. This learned behavior is stored, and repetitive acts develop the child's ability to think.

Moral Development

Moral development is not present at birth; the infant has no **conscience**, or system of values. The motivational forces guiding behavior are based on the satisfaction of needs rather than moral beliefs. Infants do what pleases them and are not aware that their acts can affect others. They react to pain and love, and judge behavior on the basis of how it affects them.

Communication

The infant at birth communicates primarily by crying to make its needs known. Studies indicate that crying has different sounds and meanings. Differences can be noted in the type and amount of crying in newborn infants. Caregiver responses can decrease or increase the amount of the infant's crying. Picking up the infant and holding, rocking, or using soothing sounds may decrease the amount of an infant's crying. Some cries may signal discomfort or pain. Determining and eliminating the source of the discomfort will lessen the crying.

Helpful Hints

- Infants who cry fretfully with their fingers in their mouth are indicating hunger.
- Infants who cry fretfully, draw their legs up in a flexed position, and pass flatus usually have colic.
- A high-pitched, shrill cry suggests injury to the central nervous system.

Before they acquire speech, infants communicate in other ways. At 2 months, the infant responds to familiar voices with pleasure and a smile. Cooing or soft throaty sounds occur at 2 months. Later, the repetition of certain sounds becomes associated with objects or persons. This is known as babbling and involves the use of consonants and vowels loosely connected together. Babbling occurs between 3 and 6 months. The sequence of sounds made in babbling is universal. Disease affecting the infant's mouth, tongue, and throat can delay babbling and language development. The 8-month-old is able to imitate simple sounds such as "da-da," much to the parents' delight. Other consonant sounds are more difficult; therefore words like "ma-ma" will be learned later. One-year-olds have an expressive vocabulary of about four to six words. They are able to understand the meaning of many more words by association with the object or by the tone of the voice. Talking and reading to infants helps increase their language comprehension and verbal ability.

All infants develop according to their own growth timetable. We have provided a rough timetable for each developmental skill for the first year of life. If marked delays in the acquisition of these skills are noted, parents should consult the physician for further evaluation. The following is a list of signs that should be discussed with the physician:

1. Moro reflex persists after 4 months.
2. Infant does not smile in response to mother's voice after 3 months.
3. Infant does not respond to loud sounds.
4. Infant does not reach or grasp by 4 months.
5. Infant still has tonic reflex after 5 months.
6. Infant cannot sit without help by 6 months.
7. Infant does not roll over in either direction by 5 months.
8. Infant does not stand by 11 months.
9. Infant cannot learn simple gestures such as waving bye-bye, shaking head yes and no by 1 year.
10. Infant cannot point to objects or pictures by 1 year.

NUTRITION

Infants' sucking, swallowing, and rooting reflexes enable them to search for and secure their food. Breast-fed and bottle-fed infants thrive equally well. Table 4–4 lists the advantages and disadvantages of both feeding methods.

Some infants are placed at the breast immediately in the delivery room. This has a number of positive effects:

1. It promotes bonding or attachment between mother and child. Early signs of bonding are evidenced by the face-to-face interaction between infant and caregiver. Other early signs include talking, smiling, and playing with the infant.

TABLE 4–4
BREAST-FEEDING VERSUS BOTTLE-FEEDING

Breast	Bottle
No preparation required	Requires preparation
Inexpensive or free	More costly
Mother must be present or milk must be expressed from breast in advance	Frees up mother's time
Milk more easily digested; causes less gastrointestinal upset; less possibility of allergic reaction	Formula not as easily digested as breast milk
Low in saturated fat	High in saturated fat
Promotes bonding with mother, but does not allow other people to feed baby	Allows father to feed and bond with baby
Baby gets immune factors from mother	Mother's diet does not affect baby
Uterus contracts; involution hastened; menstruation delayed	Amount taken at each feeding can be readily determined

2. It hastens **involution** (return of the uterus to its nonpregnant state) by stimulating uterine contractions and helping to restore muscle tone in the uterus.
3. It promotes the production of colostrum. **Colostrum** is the precursor of breast milk and is present in the mother's breast as early as the seventh month of fetal life. It contains more protein, salt, and carbohydrate but less fat than regular breast milk. In addition to these nutrients, it contains immunoglobulins to help protect the newborn until its own immune system is more developed.

Actual breast milk appears on the third day after delivery; at this time the new mother will notice that her breasts are very firm or engorged. Mothers who cannot or do not wish to breast-feed can bottle-feed with formula. Bottle-fed infants are usually given nothing by mouth for the first few hours of life. The first feeding begins with glucose and water; if this is tolerated, the infant progresses to the formula of choice. Bottle-fed newborns require a feeding every 3 to 4 hours at first and then according to their individual hunger patterns. The caregiver should never prop the bottle and leave the infant unattended. The baby should never be allowed to sleep with a bottle that contains anything other than water. Putting juice or milk in a night-time bottle can lead to **bottle-mouth syndrome: dental caries** caused by sugar in the milk or juice weakening the tooth surfaces. By 8 or 9 months, infants are ready to be weaned; that is, they can accept the cup in place of the bottle or breast. **Weaning** should be done gradually, one feeding at a time. The noon bottle is usually a good feeding to eliminate first; the baby isn't too tired or hungry to attempt learning to use the cup.

The infant's nutritional needs can be met during the entire first year by breast-feeding or by using iron-enriched formula. However, many pediatricians recommend the introduction of solid foods after the fifth month. Adding food to the baby's diet earlier is believed to add to digestive problems and possible food intolerances. One dietary concern after 5 months is that the infant's stored iron reserve is reduced. For this reason, iron-rich foods such as cereals, vegetables, and meats should be added to the diet. A daily supply of vitamin C helps to enhance the body's absorption of iron. The first solid food introduced into the infant's diet is usually rice cereal mixed with formula. It is recommended that only single-grain cereals be used at first and that egg whites, wheat, and citrus fruits not be used in the first year. These foods have been known to cause allergic reactions in many infants. It is best to introduce only one new food at a time for several days in order to detect any adverse reactions.

Helpful Hints

An approach to adverse food reactions includes the following steps:

- Alter the diet to eliminate symptoms without compromising nutrition.
- Slowly reintroduce food to see if symptoms recur; if so, remove food for 1 to 3 months.
- Discuss with physician if symptoms persist.
- Alert family members and other possible caregivers.

A general rule of thumb is to add 1 to 2 teaspoons of each new food, gradually increasing the amount to 1 tablespoon of each food item for each year. The year-old baby eats three meals per day. Table 4–5 shows a schedule of foods for the first year of life. Box 4–1 shows a sample menu for the 10- to-12-month-old child.

SLEEP AND REST

The neonate sleeps a great deal, as much as 20 out of 24 hours. The faster the rate of growth, the more sleep is required. By 1 year of age, the baby will need only about 12 hours of sleep a day. The newborn's sleep pattern is not continuous but is characterized by periods of light sleep marked by stirring movements and noises. Sleep patterns may be interrupted by discomfort and hunger.

TABLE 4–5
SCHEDULE OF FOODS, BIRTH TO 12 MONTHS

Age	Food Selections	Rationale
Birth–6 months	Breast milk or iron-fortified formula	Sucking and rooting reflexes allow infants to take in milk and formula Infants cannot accept semisolids because their tongues protrude when a spoon is put into their mouths
	Water	Small amounts may be offered in hot weather or if infant has diarrhea
5–6 months	Introduce iron-fortified instant cereal; begin with rice cereal; avoid wheat cereal for first year	Infant is now able to swallow semisolid food; cereal adds iron and vitamins A, B, and E
	Introduce plain, unsweetened fruit juices, fortified with vitamin C; dilute with equal parts water	Fruit juices provide vitamin C
7–8 months	Add plain, strained fruits and vegetables, yogurt, plain, strained meats; avoid combinations	Introduces new flavors and textures; meats provide iron, protein, and B vitamins
	Add Zwieback, toast, crackers	Teething is beginning; infant has ability and desire to chew
	Continue with iron-fortified formula, infant cereal, and fruit juices	Infant still needs iron as not yet consuming large amounts of meat
	Introduce cup to infant	Prepares infant for weaning from bottle or breast
9–10 months	Add finger foods: cooked bite-sized pieces of meat, vegetables, soft fresh or canned, unsweeted fruits, yogurt, cottage cheese; continue with iron-fortified formula, infant cereal, and fruit juices	Encourages self-feeding, and develops motor skills; introduces new textures, flavors; as formula or breast milk consumption decreases, other sources of calcium, riboflavin, and protein are needed
11–12 months	Add soft table foods: dry, unsweetened cereals, cheese slices, peanut butter, noodles	Motor skills are improving; is now ready for whole foods, which require more chewing; baby is now relying more on whole foods and less on breast milk or formula for nutrients

BOX 4–1

Sample Menu for Child at 10 to 12 Months

Breakfast
Scrambled egg yolk
or
1/2 cup cereal
1/4 cup cut-up fruit
4 to 6 oz formula

Snack
1/2 cup fruit juice or fresh fruit

Lunch
1/4 cup cooked diced poultry
1/4 cup yogurt
1/4 cut cooked diced vegetables
or
Fresh vegetables
4 to 6 oz formula

Snack
1/2 cup fruit juice or fresh fruit
1 teething biscuit or cracker

Dinner
1/4 cup noodles, pasta, rice, or potato
1/4 cup green or yellow vegetables
1/4 cup poultry or other meat, tofu, or cheese
4 to 6 oz formula

Snack
4 to 6 oz formula
(Follow with water or brush teeth before bedtime.)

A bedtime routine will help to establish a night-time sleeping pattern. This consistent approach to bedtime helps to lower anxiety and make the infant feel more secure. During the first year most infants require both a morning and afternoon nap to replenish their stamina. Table 4–6 summarizes the newborn's sleep patterns.

Sudden infant death syndrome (SIDS) is responsible for the death of about 1 out of every 500 babies, most commonly between the ages of 1 and 4 months. Although the exact cause of SIDS (also known as "crib death") is still unknown, recent research indicates an association between SIDS and sleep patterns. Death usually occurs between midnight and 6 AM. The American Academy of Pediatrics recommends as a preventative measure that healthy infants sleep on their backs and sides and not on their stomachs.

PLAY

Play is important to a child's growth, development, and socialization. The goal of play in the first year of life is nonsymbolic in that it helps the infant gain information

TABLE 4–6
NEWBORN WAKE-SLEEP PATTERNS

Type of Sleep	Activity	Duration	Intervention
Normal sleep	Eyes closed; respirations normal; occasional intermittent jerking of the body	4–5 hours per day in 20-minute cycles	Allow infant to rest; will wake if there is a sudden loud noise
Irregular sleep	Eyes closed; respiration irregular; jerky body movements; occasional groaning	12–15 hours per day in 45-minute cycles	Normal noise levels may wake child
Drowsiness	Eyes open or partly open; respiration irregular; active movement of limbs	Pattern is inconsistent	Can be easily awakened; can be removed from crib
Awake	Active movement of body and limbs; follows objects with eyes	2–3 hours per day	Position infant to interact with family members continuously; position toys so infant can play; provide other basic needs
Awake and crying	Begins with small whimpering sounds and progresses to loud crying with thrashing of limbs	1–4 hours per day	Soothe by holding and rocking; remove excess stimuli

about objects, their quality and function, and the immediate effects they produce. Different play activities help infants explore their environment. Important in the first year is choosing toys that stimulate the child's senses. In addition, it is important to select play activities that challenge and encourage musculoskeletal development. The neonate begins interacting with the environment first by following bright lights and objects.

Play during the infancy stage is solitary; that is, the infant does not require another person to play with. Infants frequently have play interactions with parents, who provide attention and stimulation. Brightly colored objects, objects that produce noise, and objects with different textures are appealing to this age group. Safety concerns must be considered when selecting playthings for the infant. All toys must be carefully inspected for sharp edges and small removable parts that may potentially be ingested or aspirated. Table 4–7 lists appropriate toys for different stages of development.

SAFETY

Because the newborn is totally helpless, the caregiver must meet all its needs for safety and protection. At this age, most injuries and many deaths are the result of preventable accidents. Safety measures include the use of approved cribs, car seats, and car beds and the prevention of drowning, suffocation, and aspiration. An infant

TABLE 4–7
PLAY AND PLAYTHINGS DURING INFANCY

Age	Play and Playthings
Newborn–1 month	Infant gazes at audible objects Dangle bells 8 to 10 inches above infant's crib Introduce rattle Play music from radio or music box
2–3 months	Infant puts hands in mouth Seat infant with an upright view of the environment Provide reach-for objects, unbreakable hand mirror
4–5 months	Infant grasps objects and brings to mouth Offer soft toys or blanket to squeeze Talk to infant and mimic sounds Offer brightly colored toys Jump or bounce on lap
6–9 months	Baby drops food or object and waits for it to be retrieved Offer squeaky toys, stuffed toys, toys with movable parts Play patta-cake and peek-a-boo Name body parts Take for rides in stroller
10–12 months	Baby stacks one block on top of another Baby scribbles with crayon on paper Baby enjoys pull toys Introduce book with animal pictures; read nursery rhymes Play simple games with large ball Blow bubbles in a cup

should never be left unattended (Fig. 4–15). Parents should be instructed to support the infant's head and neck and never to shake or jiggle the newborn. Sudden jarring or vigorous shaking of the infant's head and neck can damage the brain and cause death. As babies develop and begin to have a means of locomotion, they are at risk for different types of accidents. They must be watched constantly to prevent falls;

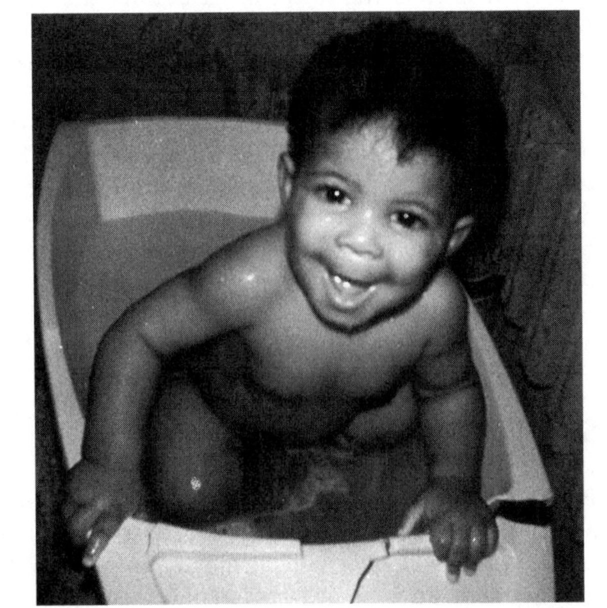

FIGURE 4–15
Infants should never be left unattended.

this is especially important when they are old enough to roll from side to side and later when they are placed in a high chair. Advise parents that, if they must turn their back even for a moment, they should secure the infant or place him or her on the floor if nothing safer is available. Babies learn and explore by putting everything in their mouths; therefore, to reduce the risk of aspiration, inspect all toys and small items and keep harmful items out of the baby's reach. It is the ultimate responsibility of the caregiver to set limits on behavior and supervise all activities to ensure safety in and around the home. Refer to Table 4–8 for information on infant safety.

Lead poisoning is a possible environmental hazard for infants and young children. Exposure may occur in older homes that often have lead plumbing or paint. If the paint is allowed to chip and peel, teething infants may pick up paint chips and chew on them. High levels of lead in the bloodstream have been associated with hyperactivity, irritability, aggression, and attention disorders. Eventually lead can be toxic to the brain.

HEALTH PROMOTION

Health promotion is aimed at assisting the infant in movement toward optimal growth and development. The nurse best accomplishes these goals by encouraging

TABLE 4–8
INFANT SAFETY

Type of Accident	Prevention
Suffocation	Use firm mattress without plastic cover. Place infant on back; do not use pillows. Keep oven and refrigerator door closed.
Falls	Provide continuous direct supervision unless infant is in crib with guardrails raised.
Choking	Keep small sharp objects out of infant's reach. Avoid such foods as nuts, hard candy, or seeds.
Poisoning	Keep medicines and household cleaning products out of reach.
Drowning	Do not leave unattended in tub or baby bath.
Burns	Provide direct continuous supervision around stove, fire, or faucet. Use flame-resistant clothing and bedding. Place guards over electrical outlets and around heating sources. Keep matches and lighters out of reach. Do not carry hot liquids while carrying or in proximity to baby.
Car accidents	Use car seat restraints that are federally approved for baby's or child's age; do not hold on lap. When using stroller, do not go behind a parked car; cross intersections with care.

education and good health practices, along with the proper use of health services, first during prenatal development and then after birth. Early assessment of the neonate can lead to early diagnosis and treatment of any abnormality. It is recommended that infants visit their health provider once a month for the first year of life. The following is a list that can be used as a guide to seeking medical attention:

- Fever >101°F
- Difficult, labored breathing
- Unexplained rash
- Absence of stools or urine
- Persistent vomiting and or diarrhea
- Extreme lethargy or hyperirritability

Caution: Acetaminophen should be used in place of aspirin for fever or discomfort because the use of aspirin for fever caused by viral infection may lead to Reye's syndrome.

Regular visits to the health center promote good health practices and health screening and allow the administration of necessary immunizations. Immunity is the body's ability to defend itself against foreign invaders such as bacteria and viruses. For the first 6 months of life, most infants have a temporary natural immunity to measles (rubella), mumps, poliomyelitis, diphtheria, and scarlet fever. At present there is no immunization against certain infections, such as the common cold. For this reason, infants must be kept away from individuals with active infectious processes. Appendix B lists the recommended schedule of immunizations. Infants should receive the immunizations according to the prescribed schedule unless they have a fever, a history of immunosuppression, or a history of allergy to the vaccine or its contents.

Helpful Hints

These are some possible side effects to routine immunization.

Minor Signs

- Localized tenderness
- Irritability
- Erythema (redness)
- Swelling at site

Major Signs

- High fever, >102°F
- Loss of consciousness
- Paralysis
- Persistent, inconsolable crying

Another area of health promotion is concerned with the family and its support system. The birth of a baby adds stress to the other family members and changes the dynamics and roles the individuals play. Chapter 2 presents a complete description of the family during the childbearing stage. All healthcare workers should be alert to signs of family stress or possible child neglect or abuse. Child abuse, which consists of physical, emotional, and sexual mistreatment, can be manifested in many different ways. It is not isolated to any one family type or social class. All signs of possible abuse should be reported at once to the proper authorities. See Table 5–7 for signs of abuse.

SUMMARY

1. Inherited characteristics are determined at the time of conception.
2. Any substance that can adversely affect the developing child is called a teratogen.
3. The period from fertilization to birth is called the prenatal period.
4. Immediately following fertilization, the new structure is called a zygote. It becomes a blastocyst at the time of implantation.
5. The developing structure is referred to as an embryo for the first few weeks and as a fetus thereafter.
6. Labor begins about 280 days after conception.
7. There are three distinct stages of labor: dilation; expulsion, or birth of the baby; and delivery of the placenta. The length of each of these stages varies with the individual.
8. Immediately after the delivery of the infant, the umbilical cord is clamped. This action ends fetal circulation and marks the infant's first breath.
9. The Apgar score is the first assessment of the newborn and is done at 1 and 5 minutes after birth. Apgar assesses infant color, reflex irritability, heart rate, respiratory rate, and muscle tone, indicating the general neurological status of the newborn.
10. The head is large in proportion to the rest of the body. The skull bones are soft to permit passage through the birth canal.
11. The anterior fontanel should close by 12 to 18 months; the posterior fontanel closes by the fourth month.
12. Average birth weight is 7.5 lb. Average newborn length is 20 in. Boys tend to be slightly larger than girls.
13. The newborn's skin is thin and delicate and varies in pigmentation. Common characteristics and skin conditions include: vernix caseosa, milia, lanugo, mongolian spots, and physiologic jaundice.
14. The newborn has the ability to swallow, digest, metabolize, and absorb nutrients. The first stool is passed within 10 hours after birth and is called meconium.
15. Primitive reflexes that are evident in the normal newborn include protective reflexes such as swallowing, gagging, sneezing, blinking, rooting, Moro, grasp, Babinski, and tonic neck reflex
16. The normal range for neonate vital signs are as follows: axillary temperature, 97.7 to 99.5°F (36.5 to 37.5°C); pulse, 120 to 140 beats per minute; blood pressure, 65/40 mm Hg (Dinamap); respirations, 30 to 60 breaths per minute.
17. Gross motor skills involve the large muscles of the extremities. Growth and development follows an orderly cephalocaudal pattern, progressing downward from the head to the feet.

18. Fine motor control of the hands and fingers follows the proximodistal directional pattern: shoulder movements are mastered before hand and finger movements.
19. According to Eriksonian theory, the infant must master the critical task of trust in order to achieve healthy psychosocial development.
20. Cognitive development is evidenced by the cause-and-effect method the infant uses to respond to its new environment.
21. Infants begin to communicate with their caregivers soon after birth by smiling and babbling. By mimicking words, infants begin to build a vocabulary. By the time babies are 12 months old, they have a vocabulary of approximately four to six words.
22. The newborn's nutritional needs can be met by either breast or formula. Generally, solids are not offered before the first 5 months in order to prevent food allergies or food intolerances. When adding new foods to an infant's diet, it is best to add one new food at a time over the course of several days.
23. The typical newborn sleep pattern includes periods of light sleep marked by stirring movements and noises. Newborns typically sleep 20 out of 24 hours.
24. SIDS, or crib death, has been associated with infant sleep patterns. This condition occurs most frequently in the first 5 months of life. To decrease the risk of SIDS, it is recommended that healthy infants be put to sleep on their backs and sides, not on their stomachs.
25. Play activities help infants explore and learn about their environment. Play during infancy is solitary. Infants need bright-colored toys free of small parts that could be accidentally ingested or aspirated.
26. Most injuries and deaths at this stage of development occur from preventable accidents. Supervision can help to decrease accidents and ensure safety.
27. Health promotion is aimed at helping infants achieve optimal growth and development. This can be accomplished through good health practices and regular medical checkups.

CRITICAL THINKING

Joyce Whitaker, age 25, has been married for 2 years and wants to have children. Joyce is attending parenting classes at the local community hospital.

1. List two critical pieces of information you would share with Joyce to help her prepare a healthy internal environment for the fetus before she becomes pregnant.
2. Once Joyce becomes pregnant, what other measures would promote the delivery of a healthy baby?

Sarah Greenoff, age 28, has been married for 8 years. Two months ago she gave birth to a baby girl named Tara. After Tara's birth, Sarah and her husband expressed concern about being effective parents. The couple was referred to the wellness clinic by the pediatric nurse practitioner.

1. Outline an assessment plan for the Greenoff family's first visit to the wellness clinic.
2. List the immunizations that are recommended for a 2-month-old infant.
3. Sarah Greenoff tells the nurse that her neighbor's baby died while asleep without any apparent warning. She is worried that the same thing might happen to Tara. Describe what information the nurse might share with Sarah to help relieve her anxiety.

Multiple-Choice Questions

1. Apgar scoring measures:
   a. The development of the fetus
   b. The newborn's neurological state
   c. The newborn's coping ability
   d. The gestational age of the neonate

2. The average length of a fully developed newborn baby is:
   a. 30 cm
   b. 40 cm
   c. 50 cm
   d. 60 cm

3. Which stool type is considered normal in breast-fed babies?
   a. Solid tan-colored stools
   b. Light seeded-mustard stools
   c. Reddish black stools
   d. Creamy white stools

4. At birth most infants are:
   a. Able to hear sounds
   b. Nonreactive to smell
   c. Unreponsive to touch
   d. Unable to show taste preference

5. Which of the following activities demonstrates that Jeremy, age 6 months, has mastered the pincer grasp?
   a. He picks up a small morsel of food.
   b. He reaches out to grasp bright-colored objects.
   c. He brings his fingers to his mouth.
   d. He rolls over from his abdomen to his back.

6. Most infants with normal motor development can sit alone at age:
   a. 4 months
   b. 5 months
   c. 6 months
   d. 7 months

7. The typical 12-month-old child will:
   a. Coo
   b. Babble
   c. Say a few words
   d. Have fluent speech

Suggested Readings

Bennett, FC, and Guralnick, MJ: Effectiveness of developmental intervention in the first five years of life. Pediatr Clin of North Am 38(6):1513–1528, 1991.

Benson, JB: The significance and development of crawling in human infancy. In Advances in Motor Development Research, ed 3. AMS Press, New York, 1990.

Biester, D: Childhood immunization: Nursing's role and responsibility. J Pediatr Nurs 7(1):65, 1992.

Blakeslee, S: In brain's early growth, timetable may be crucial. The New York Times, 29 August 1995, C1, 5.

Bock, SA, and Sampson, H: Food allergy in infancy. Pediatr Clin North Am 41(5):1047–1067, 1994.

Cox, MJ, Henderson, VK, Owen, MT, and Margand, NC: Prediction of infant-father and infant-mother attachment. Developmental Psychology 28(3):474–483, 1992.

Dilks, SA: Developmental aspects of child care. Pediatr Clin North Am 38(6):1529–1543, 1991.

Goldson, E: The affective and cognitive sequelae of child maltreatment. Pediatr Clin North Am 38(6):1481–1496, 1991.

Greenspan, SI: Clinical assessment of milestones in infancy and early childhood. Pediatr Clin North Am 38(6):1371–1385, 1991.

Howard, BJ: Discipline in early childhood. Pediatr Clin North Am 38(6):1351–1369, 1991.

Hubbard, F, and Ijzendoorn, M: Maternal unresponsiveness and infant crying across the first 9 months: A naturalistic longitudinal study. Infant Behavior and Development 14:299–312, 1991.

Juneau, PS: Essentials of Maternity Nursing. FA Davis, Philadelphia, 1985.

Lawrence, PB: Breast milk, best source of nutrition for term and preterm infants. Pediatr Clin North Am 41(5):925–941, 1994.

Moskowitz, B: The acquisition of language. Science America 239(5):89–109, 1978.
Redel, C, and Shulman, R: Controversies in the composition of infant formulas. Pediatr Clin North Am 41(5):909–924, 1994.
Sewell, K, and Gaines, S: A developmental approach to childhood safety education. Pediatric Nursing 19(5):464–466, 1993.
Steinschneider, G, Glassman, M, and Winn, K: Sudden infant death syndrome prevention and an understanding of selected clinical issues. Pediatr Clin North Am 41(5):967–989, 1994.
Stevenson, RD, and Allaire, JH: The development of normal feeding and swallowing. Pediatr Clin North Am 38(6):1439–1453, 1991.
Vaughn, B: Attachment security and temperament in infancy and early childhood. Developmental Psychology 28(3):463–473, 1992.
Whaley, L, and Wong, D: Nursing Care of Infants and Children. Mosby-Year Book, St. Louis, 1991.

Chapter 5

Chapter Outline

Toddlerhood

Key Words

acuity
ambivalence
amblyopia
autonomy
egocentric
eustachian tube
lordosis
negativistic behavior
ossification
parallel play
regression
ritualistic behavior
separation anxiety
sibling rivalry
socialization

Learning Objectives

At the end of this chapter, you should be able to:

- Describe the main physical characteristics common to toddlers.
- Name three developmental skills that the toddler can master independently.
- Describe the psychosocial task of the toddler as outlined by Erikson.
- List one method of discipline useful in resolving conflicts during this stage.
- Describe the stage of cognitive development for the toddler as presented by Piaget.

- List two factors that help toddlers develop language skills.
- List three feeding recommendations for parents of toddlers.
- Describe the type of play typical for toddlers.
- Name five common safety hazards for this period of development.

The toddler period usually refers to the period from 1 to 3 years of age. After the fast growth spurts of infancy, the growth rate of the toddler is slow and steady. Many new skills are being developed, including both fine and gross motor skills related to dressing, feeding, toileting, and walking. Another accomplishment during this period is related to language development. These newly acquired skills serve to strengthen the toddler's newfound autonomy.

PHYSICAL CHARACTERISTICS

Height and Weight

The toddler usually grows an average of 3 inches (7.5 cm) per year. The average height of the child at 2 years is 34 inches (86.6 cm); at 3 years it is 37.25 inches (95 cm). The toddler gains an average of 4 to 6 lb (1.8 to 2.7 kg) per year during this period. By age 2, the toddler averages 27 lb (12 kg). At 3 years, the child usually weighs 32 lb (14.6 kg).

Body Proportions

The child's extremities grow much faster than the trunk, resulting in a more proportionate appearance for the body as a whole. The typical 2-year-old has a potbellied appearance—a large belly and an exaggerated lumbar curve, known as **lordosis** (Fig. 5–1). By the end of the third year, the child is taller and more slender, with stronger abdominal muscles and a more erect posture. Head growth slows down in comparison to the rate of growth in the body and extremities.

Face and Teeth

The face and jaw increase in size to permit room for more teeth. At 2½ years the child can be expected to have 20 teeth, a complete set of deciduous or primary teeth. Children at this age should visit the dentist for a preliminary dental examination and dental supervision (Fig. 5–2). Parents should discuss with their dentist the possible need for fluoride treatments, depending on their local water supply. Parents should also help the child learn self-care to maintain oral hygiene. Toddlers should be supervised while brushing their teeth at least twice a day.

Bone Development

Like the child's general growth, bone growth and development are greatest in the first year and are followed by a gradual slowing down. As the child grows, the bones increase in density and hardness. Cartilage is gradually replaced with bone tissue. This process, known as **ossification**, will not be completed until puberty. The hardening of the soft spongy tissues is gradual and occurs at different rates for different parts of the body. For example, by 18 months the toddler's anterior fontanel will be closed, although other bones still remain soft and pliable. This explains why

FIGURE 5–1
Weak abdominal muscles give the toddler a potbellied appearance.

infants and young children appear more flexible and can bend and put their toes into their mouths. In fact, this also explains why some young children develop a type of bone fracture known as a "greenstick fracture." In this fracture the bone is angulated beyond the limits of normal bending, similar to the bend in a green, unripe stick.

Sensory Development

The toddler's visual **acuity** changes gradually. The eye muscles strengthen, which further develops binocular vision. The toddler's vision may be 20/40—and better when large objects are placed at a distance of 6 feet. Visual acuity will improve to 20/20 by the end of this stage. Depth perception improves as the child enters toddlerhood, but will not be fully developed until later, during the preschool period.

Some children have a condition called "lazy eye," or **amblyopia**. Children with amblyopia have double vision, but have no way of knowing it because they have nothing to compare it to. It is imperative that toddlers undergo vision screening to

FIGURE 5–2
Dental care and regular visits to the dentist should begin at about 2½ years.

help detect this condition. The current treatment for amblyopia involves patching the stronger eye to force the child to use the weak eye. Corrective lenses and exercises also help correct this condition. Untreated amblyopia can lead to blindness in the affected eye. Strabismus, or crossing of the eyes, may be seen from time to time; if it persists, professional attention is indicated.

Hearing in the toddler is fully developed. Routine physical examinations should include periodic hearing tests to detect any changes from the norm. Frequent monitoring should be done in a child with delayed speech or with repeated ear, nose, and throat infections. One structure of the ear, called the **eustachian tube**, connects the middle ear to the oral pharynx. This structure is shorter and narrower than in older children, allowing easy passage of microorganisms from the upper respiratory tract to the middle ear. This accounts for a higher incidence of ear infections in the young child compared to the older child. Children with a history of ear infections can be at risk for hearing loss. A common sign of an ear infection in young children is unrelenting crying and rubbing or pulling of the affected earlobe.

VITAL SIGNS

The toddler's temperature-regulating mechanism is more stable and fully developed than that of the infant. For this reason, the toddler is not so sensitive to environmental changes. The toddler's body temperature is maintained at the normal range of 98 to 99°F (36.6 to 37.2°C). The heart rate slows down because the heart is larger and more efficient. The average pulse rate is between 90 and 120 beats per minute. Respirations slow down to 20 to 30 breaths per minute as a result of increased lung efficiency and capacity. At this stage the average blood pressure measurement is 99/64 mm Hg.

DEVELOPMENTAL MILESTONES

Motor Development

The acquisition of skills in the toddler is based on the further development and refining of the crude gross and fine motor abilities achieved during infancy. The motivational force behind the development of these skills is rooted in the child's search for independence. By the end of toddlerhood, the child will have developed skills related to independent functioning, including walking, eating, toileting, dressing, and language use.

Gross Motor Skills

Gross motor skills depend on growth and maturation of the muscles, bones, and nerves. Until a state of readiness is reached, teaching the child developmental skills such as walking, skipping, or hopping is of little value. Once readiness occurs, the child needs to be given ample time to practice and master each new skill attempted.

Walking, first seen with an unsteady gait, may begin for some children by their first birthday and for others several months later. Typical skills of the average 15-month-old include walking alone without assistance, limited balance, and creeping up stairs. The 18-month-old typically can walk up stairs with both feet and sit down on a chair. The 18-month-old runs clumsily, which results in frequent falling. The 2-year-old child can climb the stairs alone using two feet on each step, run with a wide stance, and kick a large ball without losing balance and falling. At this age, the child can walk down stairs with assistance, jump in place with both feet, and sit on a chair independently. The 3-year-old can hop, stand on one foot, and walk a few steps on tiptoe.

Fine Motor Skills

Fine motor skills include those centered on self-feeding, dressing, and play. By 15 months, a toddler can more deftly grasp a spoon and insert it into a dish, but will continue to invert and spill the contents until the end of the second year. Most toddlers are fascinated with dining utensils, but the majority use their fingers and prefer finger foods.

The 1-year-old can usually remove his or her socks, shoes, hat, and mittens. By the end of the second year, the child can remove all of his or her clothing and will attempt to put some items back on, but only if the child chooses to. Toddlers have enough fine motor dexterity to allow them to wash themselves. This is often seen as a pleasurable activity, but they frequently wash only the face and stomach while ignoring the rest of the body.

The 3-year-old's fine motor coordination improves to permit the child to now hold a crayon with the fingers instead of using the fist. The 3-year-old can control drawing to include both vertical and circular strokes. There is slight evidence of hand preference during the first year of life. Usually infants reach for objects with both hands or the hand closest to the object. Right- or left-hand dominance is evident by 15 months. Table 5–1 lists developmental milestones for a 3-year-old.

TABLE 5–1
DEVELOPMENTAL MILESTONES FOR THE 3-YEAR-OLD

Gross motor	Balances on one foot Jumps on both feet Walks up steps using both feet Runs Rides a tricycle
Fine motor	Puts simple puzzles together Can build a tower of blocks Copies a circle or vertical line Can turn knobs and open lids
Psychosocial	Attached to mother but can tolerate short separations Dresses and undresses self Possessive of own property Nearly fully toilet-trained
Cognitive	Searches and finds toys Able to locate body parts Knows relationship between things and persons Gives full name
Language	Uses words and gestures to indicate needs Uses two-word sentences Imitates sounds and words Sings simple songs Has vocabulary of 1000 words

Toilet Training

Toilet training is often more important to the parents than to the young child. Successful toilet training depends on a certain degree of maturity in the toddler's muscles, including sphincter control and maturation of the sensory centers of the brain. Furthermore, toddlers must develop a system of communication that allows them to alert parents to their needs. The child uses either gestures or words to convey his or her need for toileting. Toddlers will learn bowel control before mastering bladder control. Usually children do not have this control until the second year, after walking for several months. Most children achieve daytime dryness long before night-time dryness. Children of this age need help undressing and with the whole toileting process. By the time the child is 3½ years old, he or she is usually bladder-trained. Changes in schedules, emotional stress, fatigue, or illness can often cause setbacks in toilet training. It is best to expect accidents at times when children are engrossed in play or simply miss the signals. These accidents should be handled in a matter-of-fact manner, without punishment, to help build the child's self-esteem.

Psychosocial Development

Autonomy

According to Erik Erikson, **autonomy**, or independence, is a major psychosocial task of the toddler (see Chap. 3). In particular, toddlers are trying to master independence in their daily activities, such as toileting, dressing, feeding, and taking care of their belongings. Encouraging children to make simple decisions fosters a sense of independence. Freedom of choice, however, often sets the stage for conflict of interest between parent and child. That is, fostering autonomy does not preclude parental guidance. Many activities, particularly those concerning the child's safety, such as playing in the street, are nonnegotiable. How these conflicts, as well as mishaps and successes, are handled is critical to the toddler's developing self-esteem. If the toddler is punished for accidents and made to feel worthless, he or she develops a sense of shame and doubt. For example, if a toddler has an accident and soils his or her clothing, parents should not get angry and scold the child. They should try to have a change of clothing handy and treat the accident as a minor event: "You'll do better next time." Even with nurturing guidance, toddlers often develop conflicting emotions or feelings of **ambivalence** as they are expected to learn independent behaviors. For instance, toddlers may experience feelings of both love and hate for their caregivers when being reprimanded or disciplined. Toddlers may get angry with parents and say, "I hate you" and still want to be held and comforted.

Parents should be careful not to take over dressing the child who is attempting to carry out this skill. Although it is quicker for the parents to dress the toddler, it is better to support development of the child's independence by allowing the child to practice these skills. Box 5–1 lists principles for understanding behavior.

BOX 5–1

Principles for Understanding Behavior

1. Behavior must be understood in the context in which it occurs by looking at all related factors.
2. All behaviors have objectives and serve a purpose.
3. The child's self-concept influences his or her behavior.
4. Behavior helps the child maintain psychological equilibrium.
5. The child's perception and interpretation of behavior influence his or her actions.

Discipline

Toddlers need discipline because they do not have enough information to understand what is acceptable or unacceptable behavior. A simple, direct "No" followed by some diversion will help lay the foundation for learning impulse control. It is crucial that caregivers be consistent and repeatedly reinforce limitations. Discipline should not deny the child freedom but give the child a greater opportunity to explore and learn within safe limits. Discipline should guide, correct, strengthen, and improve the child's choices. Nonnegotiable issues include items such as not hurting themselves or others, not destroying property, and not placing themselves in unsafe conditions such as running into the street. With these non-negotiable issues the parents should give clear, simple instructions, such as, "Bobby is a friend to play with, not to hit!"

Sometimes discipline triggers temper tantrums or rebellious behavior. This **negativistic behavior** occurs as a result of frustration that the child encounters when his or her needs or wants are not met immediately (Fig. 5–3). The toddler is

FIGURE 5–3
Negativistic behavior occurs as a result of the child's frustration.

eager to take control and be independent beyond what skill or judgment allows. Toddlers have a limited vocabulary; this makes it difficult for them to express their feelings and may result in outbursts of kicking, screaming, and breath holding when they cannot have their way. Temper tantrums are commonly seen between ages 2 and 3 and diminish in intensity and frequency by age 4 to 5. To avoid conflicts, parents can place less emphasis on minor issues and allow the child to make some choices. For example, the parents may find it better not to rush or hurry the dawdling child, or they may allow the child to postpone dressing until after breakfast. Giving choices when possible may help to reduce the number of conflicts and temper tantrums. If a tantrum occurs, the parents should ensure the child's safety and, if possible, leave the child in his or her room or limit the number of onlookers.

Helpful Hints: Temper Tantrums

- Ensure the safety of the child.
- Stand back, take a deep breath, and wait a few seconds before responding.
- Take control of the environment and remove the child to a neutral site.
- Ignore the tantrum without ignoring the child.
- Use a controlled tone of voice; never scream.
- Acknowledge the child's upset feelings.
- Avoid reasoning when the child is out of control.
- Don't worry about what people around you think.
- Be consistent and firm in setting limits without anger.

Another intervention that may be used to resolve a conflict of wills is the concept of "time out." The child is usually removed from the center of activity to a quiet place where he or she can regain some control. Time out should be immediate and used only for a few minutes. Following the time-out period, the parent and child should talk about the events leading up to the conflict and possible solutions. This teaches the child to talk about what he or she is feeling and helps the child learn alternative solutions to problems.

It is very common for 2-year-olds to demand that things go their way. It is usually best not to share plans in advance of an expected event. If the parent or caregiver promises to take the child to the park on the following day, the child may demand fulfillment of these plans regardless of bad weather or other happenings. Sometimes parents need to ignore attention-seeking behavior when it does not put the child at risk or in danger. Caregivers should remember to offer praise and positive reinforcement for desired behaviors. See Box 5–2 for tips on discipline.

BOX 5–2

Tips on Discipline

- Try to understand the reason for the misbehavior.
- Respect the child as a person.
- Be firm but kind.
- Be patient.
- Reward and praise often.
- Encourage open expression of feelings.
- Ignore negative behavior as safety allows.
- Provide a healthy environment.
- Listen and be attentive.
- Encourage independence.
- Avoid pity.
- Maintain control of emotions.
- Allow for trial and error.
- Reinforce consequences.
- Use familiar routines when possible.
- Model desired behaviors.
- Offer choices.

Special Psychosocial Concerns

Toddlers are affected by what is called **separation anxiety**. As they become more independent, they can tolerate only brief periods of separation from their parents. Children in this age group are naturally warm and affectionate (Fig. 5–4) but still somewhat fearful of strangers unless they are accompanied by a family member. Parents should be honest about leaving or going out without the child and telling them when they will return. This helps to reinforce the idea that they will come back as they said they would.

FIGURE 5–4
Toddlers are warm and affectionate.

Toddlers sometimes use certain "comfort items" to decrease their anxieties. These items are often blankets, soft toys, or other common household items. These transitional objects, as they are sometimes called, are important to the child but often a concern to the parents. Children should not be expected to give up their transitional objects all at once. Significant changes in the child's life, such as arrival of a new sibling, a move, or the onset of preschool or day care, may cause stress and result in the need for the comfort item.

Helpful Hints: Giving up Comfort Items

- Recognize the transitional object as a part of the child's journey toward independence.
- Try to offer a substitute item rather than insisting on removing the object.
- Allow enough time for the child to reach a state of readiness.

The birth of another child often creates **sibling rivalry** or feelings of jealousy and insecurity in the toddler. It is difficult for a 2- or 3-year-old child to share time, attention, and parental affection with a brother or sister. Angry outbursts or regressive behavior may be seen in the toddler trying to deal with changes in the family brought about by the birth of a new baby.

Regression, a return to an earlier form of behavior with which the child felt comfortable and secure, can occur at any stressful time. It is not uncommon for the toilet-trained child to regress and have accidents, particularly following an illness or separation from parents. When this form of regression occurs, parents should minimize the significance and be confident that, once the stressful period is over, the child will return to normal behavior patterns. Sometimes after the birth of a new baby, the toddler may regress and want to use a bottle or be carried around like the newborn child. Parents should expect some regression and plan for special time alone with the older child to make him or her feel as important as the new baby.

Cognitive Development

Cognition continues to develop by trial-and-error. At 2 years of age the child begins what Piaget describes as preoperational thought (see Chap. 3). The toddler's problem-solving abilities are limited. For example, the child may correctly identify a garbage receptacle but indiscriminately throw everything into the receptacle. The child also responds to the total situation rather than to a part. In other words, when objects and things have common elements, the child will respond to them as if they were all the same. As memory begins to develop, many behaviors are imitated. The toddler's interpretation of new experiences is based on memory of previous happenings. For example, the toddler will recognize familiar objects and people and respond to them with pleasure but may demonstrate a fear of unfamiliar things and strangers.

Toddlers begin to experiment by trying out new ideas or actions. As object permanence develops, thinking follows a simple and direct pattern. At this stage, the child no longer believes things will disappear if they cannot be seen. The child uses the further development of memory to create mental images. These mental images have a magical quality and are usually incorporated into the toddler's play. Events are seen as having a simplistic causal relationship. For example, toddlers believe that their own feelings can directly affect events. Some children are concerned that their angry thoughts may cause bad things to happen. Toddlers have an **egocentric** view of the world; that is, they cannot sense the world from any point of view other than their own. For example, a child grabs a toy from another child but cannot understand that this may hurt the other child's feelings. It is best for caregivers to simply explain that this behavior is not acceptable.

Time is still incompletely understood or can be interpreted by events within the child's own frame of reference. It is best to avoid using words like "tomorrow," "yesterday," or "next week." It is clearer to the child if the speaker uses a familiar event to relate to a particular happening; for example, "We will go out to the park after your lunch." This is more meaningful than saying, "We will be going to the park in the afternoon."

Moral Development

As they grow, children learn their moral values based on their parents' moral codes and by imitating parental behavior and teachings. Parents begin to teach toddlers what is right and wrong. For example, they are taught that it is wrong to stand in the car and that instead they must always be buckled and seated in their safety seat. Parents must use their own seat belts when driving or riding in a car. This sets a good example for children and helps to teach them right from wrong. Repeated instructions and consistency reinforce moral decisions. All caregivers should work together as a team to help instill the same principles. Respect between the parent and child teaches the toddler that justice is reciprocal. Reasonable discipline that draws on the respect for others is an integral part of the child's own moral development. Learning socially acceptable behaviors is a long, slow process that begins in this stage and extends through adolescence. Refer to Box 5–3 for tips on promoting self-esteem.

BOX 5–3

Promoting Self-Esteem

- Attend to needs immediately.
- Spend special time with the child.
- Ignore minor mishaps.
- Listen attentively.
- Convey positive regard.
- Label behavior, not the child.
- Give positive feedback.
- Be congruent with communication.
- If indicated, offer an apology.

Communication

Language acquisition is automatic and spontaneous. Language skills are enhanced with practice. Encouraging speech and reading to children help build language skills. The toddler's language is based on symbolic function or memory. This means that their words not only name things but show they understand their meaning. In other words, if the child uses the word "potty," this represents the process of toileting. Cognitive development and imitation play have important roles in early language acquisition. Children understand what is said to them before they are able to put their thoughts into words.

Early on, sentence structure may not be correct, nor is pronunciation clear, but toddlers are still able to complete their intended message. Sometimes only the immediate caregivers can understand the child's language. The first sentences consist of a noun or a verb. Toddlers quickly develop the ability to use nouns or verbs in a two-word sentence. To make the two-word sentence clearer, most toddlers use hand gestures to support the meaning of their words.

Most 2-year-olds are able to use words to represent their actions and make their needs known. For example, if a toddler is thirsty, he or she may say "juice" repeatedly until the need is met. When the toddler learns the use of the word "why?" it is often used to challenge adults and keep them talking. This dialogue ultimately helps the child to learn more about the world. By 2½ to 3 years of age, the toddler begins to use short three-word sentences. Toddlers often confuse the pronouns "I" and "me." "Mine" becomes a part of the child's vocabulary rapidly thereafter, because at this time the child begins to show awareness of ownership. Everything becomes "mine." From 50 words at age 2, the child moves rapidly to a vocabulary of

1000 words by age 3. The 3-year-old can put a noun and verb together to create a short sentence such as "I go" or "Give it to me."

Young children living in a bilingual family can learn more than one language at the same time. Bilingualism is possible for the toddler if both languages are used in the home. When one language is used at home and another at play group or nursery school, it is much more difficult for the child to learn the second language.

NUTRITION

The toddler needs to establish good eating patterns because the eating habits taught at this stage will be lifelong. Because toddlers' eating habits are easily influenced by the eating preferences of older siblings and parents, other family members need to model good eating practices. It is especially important that caregivers provide them with the appropriate amounts of foods from the various categories in the food pyramid.

Toddlers particularly need foods that allow muscle development and mineralization of the bones; that is, foods containing adequate protein, calcium, iron, phosphorous, and vitamins. One daily serving of green leafy or yellow vegetables, along with fruits, will provide the needed vitamin C. Foods with adequate fiber should be encouraged. Iron-rich foods such as cereals, meats, and fruits should be provided. Because most children continue to like milk, the milk intake must be monitored and limited to 1 quart per day. The need for milk will decrease as the diet of solid foods increases. By age 2 to 3 years the child should be eating the same foods as the rest of the family.

Most toddlers need about 1300 calories a day, although the amount of food needed varies greatly, depending on the child and his or her activity level. The child who is very active needs more calories than the child who is sedentary.

Allowing a toddler to select junk food, including food high in sugar, is detrimental to the child's nutritional state. In addition to meals, snacks need to be nutritious. Candy and other concentrated sweets should be avoided or given very sparingly. Snacks should be tallied in daily considerations of calories, proteins, and other vital nutrients.

Caregivers can comply with many of the toddler's food preferences and still offer nutritious foods. Most toddlers prefer plain foods to mixtures. They will learn about textures from a variety of foods offered. Foods that are easy to manipulate and chew are among toddlers' favorites. For example, hand-held sandwiches, bite-size pieces of meats, pizza, pasta, and fruits are some of their favorite foods. Initially young toddlers bite and chew with their front teeth; however, as the back teeth (molars) appear, they will begin to use the back of the mouth. Because toddlers have their full set of deciduous teeth by 2½ years of age, they are able to chew and swallow all sorts of table foods.

It is not uncommon for children of this age to develop **ritualistic behavior** in relation to eating. For example, toddlers sometimes prefer using the same plate or cup at each meal. Their ritualistic preferences may be upsetting to caregivers, who should remember that such behavior is typical for many toddlers. In addition, toddlers may develop food fads or habits in which they go for a time not eating or eating only small amounts every day. Common reasons for toddlers not eating include excitement or distraction, exhaustion, illness, lack of hunger, and attention seeking. Eating the same foods every day may be boring to an adult, it but will not be detrimental to the child's nutrition if the foods selected contain the appropriate nutrients. And a food fad is likely to disappear just as suddenly as it began.

Toddlers also like consistency and familiar routines at mealtimes. These qualities foster good eating habits; thus meals should be at the same time each day. Because toddlers' stomach capacity is small, caregivers should plan for three small meals along with three nutritious snacks. See Box 5–4 for tips for nutritious snacks.

BOX 5–4

Tips for Nutritious Snacks

- Snacks can be an excellent means of providing additional calories, proteins, and other vital nutrients.
- Avoid offering snacks immediately before mealtimes.
- Avoid the use of concentrated sweets.
- Make healthy choices: cheese cubes, fresh fruits, raw vegetables, milk, crackers, dried cereals, dried fruits, peanut butter on bread or crackers, plain low-fat yogurt, etc.

Mealtimes should be used to promote family time together and **socialization** of the toddler, including promotion of the child's autonomy in self-feeding. It is best at this stage to offer simple choices, such as "Would you like cereal or toast for breakfast?" This is better than offering vague choices, such as "What do you want for lunch?" Children sometimes test their autonomy by refusing to eat. Toddlers may also refuse to eat because their appetite fluctuates and they may not be hungry at mealtime. Caregivers should be careful not to offer snacks too close to mealtimes. Toddlers may also be too tired, too excited, or too distracted to eat. Sometimes refusal to eat is an attention-seeking behavior. It is best to ignore most refusals because they are usually short-lived: the child will eat when he or she is hungry. At times, toddlers will play with their food or dawdle. They need to be given adequate time to finish eating but not so much time that one meal runs into the next.

Caregivers should give as little attention to a toddler's negative behaviors at mealtimes as possible. To promote good eating habits and socialization of the child, mealtime should not be stressful. Parents should expect that accidents and spills will happen and react matter-of-factly to unintentional happenings. The toddler can be taught simple table manners: how to use the correct eating utensils, for example. Using the proper utensil also enhances fine motor coordination. Eating utensils can be of normal size or child-size, if desired. Caregivers should provide positive reinforcement for desired mealtime behaviors. Sometimes setting the table with a special cloth or with the "fancy" dinnerware helps to show toddlers that they are growing up and able to use "the grown-up stuff." See Box 5–5 for tips to foster good eating habits.

SLEEP AND REST

Toddlers sleep less than infants and often resist sleep because they want to play and be involved in adult activities. Short nap periods during the day will help prevent toddlers from becoming overtired. Bedtime should involve some kind of a ritual such

BOX 5–5

Fostering Proper Eating Habits

- Encourage the tasting of new foods.
- Introduce new foods in small amounts along with regular foods.
- Use child-size portions.
- Present colorful foods of different textures.
- Eat with the child.
- Never force a child to eat.
- Provide a comfortable atmosphere.
- Minimize confusion at mealtimes; for example, turn television off.
- Recognize that accidents or spills will occur.

as reading to the child and allowing the child to have some kind of comfort toy such as a teddy bear or blanket. Ritualistic behavior or habitual acts surrounding bedtime practices can establish a familiar routine for the child to follow. Such rituals help reduce anxiety and give the child a sense of security.

Sleep disturbances caused by nightmares are not unusual, causing the child to wake up in the night frightened. The child usually resists sleep after a bad dream because of fear. The events surrounding the nightmares usually appear real and lifelike, and the child will need to be comforted. The period of comfort should be brief, and it is usually recommended that parents comfort the child in the child's room rather than taking the child to their own bed. Toddlers should understand that each person in the family has his or her own space for sleeping. Taking the toddler into the parents' bed may set up a comfortable habit that later can be hard to undo. Other factors that can produce sleep disturbance in the toddler include fear of separation from the parent, illness, and physical exhaustion.

Helpful Hints: Resolving Sleep Problems

- Set a consistent routine and remember that you cannot force a child to sleep.
- Encourage quiet rituals before bedtime.
- Use a night light if the darkened room frightens the child.
- Remember that not all children require the same amount of sleep.

PLAY

Play is a very important activity in the toddler years and is the major means by which children continue to explore and understand the world around them. At first, play mimics the activities performed by others around the toddler, such as talking on the telephone. This type of mimicking is not only pleasurable but helps the toddler try out adult roles. By 2 years of age the goal of play appears to be symbolic rather than the nonsymbolic play of infants. Nonsymbolic play is demonstrated when the young child squeezes a soft ball. Piaget described symbolic play as the emergence of make-belief and pretense. Objects become the symbol or represent something else remotely similar (Fig. 5–5). Symbolic play is at its peak between the ages of 2 and 4 years. The importance of symbolic play is believed to be that it serves to help children explore different possiblities, control aggression through fantasy, and pretend. Examples of symbolic play can be blocks used to build a building or city. Other examples of symbolic play occur when the child first pretends to drink

FIGURE 5–5
Different forms of symbolic play during the toddler years. *(A)* Dressing up. *(B)* Playing with dolls. *(C)* Cleaning house. *(D)* Talking on the telephone.

from a cup and later pretends to feed a doll. Parents have the fundamental responsibility in guiding their child's play by modeling the desired activity.

Play is good in that it helps in ego development, cognition, and socialization. Many 2- and 3-year-olds are enrolled in some type of structured play group. Even though they cannot always interact with other children at play, they benefit from being in the presence of children of the same age group learning socializing skills (Fig. 5–6). Play groups may also be beneficial for the mother or caregiver, allowing time for other siblings or themselves. Children with working parents may develop similar social skills in a day-care setting. Although children are encouraged to play with other children, many confrontations occur over playthings. Toddlers usually prefer **parallel play**, in which they play alongside other children without interactions.

As toddlers develop language proficiency and the capacity to think, they are capable of incorporating their elaborate imagination into their play activities. Many 3-year-olds, for instance, develop imaginary playmates. Girls have imaginary playmates more often than boys. These imaginary friends are often blamed by the child when accidents or other mistakes occur. This type of play is a very normal part of development and should be treated with sensitivity by caregivers.

The selection of playthings is one of the many critical decisions parents have to make (Box 5–6). Toys foster fine and gross motor development as well as entertain. Push toys, riding toys, swings, and pots and pans assist with gross motor

FIGURE 5–6
Toddlers need the presence of children of a similar age.

BOX 5–6

Play and Playthings for Young Toddlers

Parallel play: Plays alongside other children but does not interact or share

Playthings: Tricycles, swings, climbers, rocking horse, color cubes, paint and brushes, and simple musical instruments such as drums or bells

development. Finger painting, drawing, puzzles, and building blocks strengthen fine motor development. Some 2-year-olds like to unscrew tops from bottles, open boxes and containers, turn pages in a book, and cut with scissors (Fig. 5–7). Toddlers should use safety scissors under grown-up supervision. Many of the objects used in these favorite activities are readily found in the home.

SAFETY

Because of their natural curiosity and explorative behavior, toddlers are prone to accidents, and these become the most frequent reason for medical services. Accidents are the leading cause of death in children of this age. Most accidents are preventable and require parental education. The toddler requires constant supervision, as there are many hazards both inside and outside the home. Toddlers are incapable of recognizing any danger or threat to their safety; this, along with their curiosity, places them at great risk for injury (Fig. 5–8).

Recent statistics indicate that half of all accidents resulting in death or serious injury occurring at this age are motor vehicle accidents. Car safety seats and restraints are now under state regulation and strictly mandated to reduce the numbers of injuries and deaths. Children must be taught what is expected of them with regard to car safety. For example, the child must know that he or she is expected to be seated in the proper seat and restrained when in any vehicle. Windows and door

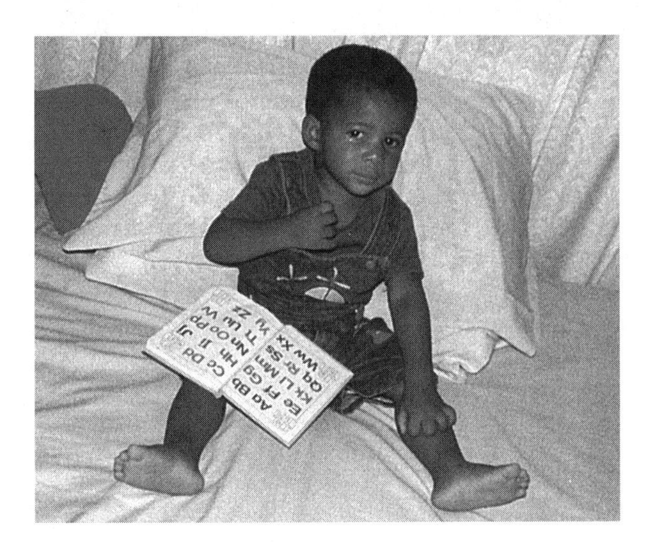

FIGURE 5–7
Toddlers like to turn the pages of a book and to have stories read to them.

FIGURE 5–8
Curious toddlers are often at risk of falling and other accidents.

handles should be off-limits to all young children. Car manufacturers offer the consumer the choice of purchasing added safety features such as window and door locks. A consistent approach to rules pertaining to car safety, or safety in general, will better ensure compliance.

Play areas should be carefully selected to provide child safety. Children should not play near roads or in active driveways. Continuous watchful supervision is necessary to prevent accidents. No child of this age can be trusted to remember all the safety rules or be able to recognize potential dangers.

The ingestion of poisons is another safety hazard as the toddler begins to climb and open drawers and closets. Special care must be taken in storing household cleaning agents, garden products, car products, and prescription and nonprescription drugs. These agents should be stored on a high shelf, out of the reach of children and securely locked. Parents should never refer to medicine as "candy" or say that it tastes "good"; otherwise, toddlers may be tempted by it. Parents should make it a point not to take their own medications in front of their children in order not to encourage children to copy their behavior. All caregivers should have the poison control number readily available and two doses of ipecac (a drug used to induce vomiting) on hand in the event of accidental poisoning.

Injuries resulting from burns rank second to motor vehicle accidents. Safety in the kitchen includes removing stove knobs and turning pot handles toward the back of the stove to prevent spills. Electrical sockets should be covered with safety caps when not in use to prevent shocks. Children are at risk of burns from hot liquid spills or bath water or from playing with matches.

Choking or aspiration is a continued concern at this age, particularly because the toddler often eats food on the run. Certain food items have been identified as frequent offenders. They include hard candies, popcorn, fruit pits or seeds, and large pieces of meat. Another potential safety hazard is toys with small removable parts that can be placed in the child's mouth and either swallowed or aspirated.

This kind of toy should not be given to young children. Suffocation can be avoided by advising parents not to use plastic bag coverings on beds or furniture. Balloons have been implicated in many toddler deaths: while attempting to blow up a balloon, toddlers may accidentally inhale the deflated balloon into the windpipe, obstructing the airway.

Drowning accounts for high numbers of deaths in this age group. This can occur in the bathtub, swimming pool, or other body of water. Drowning may occur in only a few inches of water. Baths should be supervised at all times. Adult supervision and teaching children to swim can make water a safe, fun activity (Fig. 5–9).

HEALTH PROMOTION

Regular physical examinations should be scheduled for toddlers. These exams include monitoring the child's growth patterns, health screening, identification and correction of any deviation, education, and disease prevention. Visits to the health provider should be scheduled when the child is 18, 24, and 36 months. A history and physical examination is done at each yearly visit; it should include an assessment of the child's growth and development. At the 24-month checkup, the child will need to be screened for tuberculosis. To strengthen parenting skills, child-rearing classes may be indicated. Toddlers should be immunized against measles, mumps, and rubella vaccine at 12 to 15 months of age. The varicella vaccine is currently recommended for the 12- to 18-month-old toddler. The complete schedule of recommended immunizations is found in Appendix B.

Healthcare workers, caregivers, and teachers should be alert to signs of maltreatment or abuse. As with infants, abuse of any kind can occur in any family type and at any level of social standing. Child abuse can include physical, emotional, and sexual maltreatment. See Table 5–2 for the common signs of child abuse.

FIGURE 5–9
Learning to swim at an early age is important for water safety.

TABLE 5–2
SIGNS OF CHILD ABUSE

Physical abuse	Bruises, welts (may be at different stages of healing) Signs of multiple fracture at different stages of healing Lacerations or tears Cigarette or immersion burns on extremities or buttocks Electrical appliance imprint Head injuries Swollen, blackened eyes
Sexual abuse	Difficulty walking or sitting Bruises or bleeding from genitalia Torn, stained underclothes Recurrent urinary tract infections
Psychological/affective abuse	Excessive anger, aggression Poor peer relationships Negativism, loss of pleasure Low self-esteem, lack of trust Developmental delays Withdrawn behavior, regression

SUMMARY

1. Toddlerhood refers to the period of development from 12 months to 3 years of age.
2. Growth rates slow down compared to the rate of growth during infancy. Growth during this period results in the body appearing more proportionate, with a taller, more slender look.
3. Bone development continues, with a gradual hardening or ossification.
4. By $2\frac{1}{2}$ years of age, toddlers usually have a complete set of deciduous teeth. Children at this age must visit the dentist for dental examination and treatment to help ensure healthy teeth later on in life.
5. Gradually visual acuity will improve and can be enhanced with use of large objects held at close range. Hearing is fully developed and should be tested during periodic examinations.
6. Heart rate, respiratory rate, and blood pressure readings will slow down during toddler period.
7. Gross and fine motor skills are developing further.
8. By the end of toddlerhood, the child will gain skills relating to walking, eating, toileting, dressing, and communicating independently.
9. Development of gross motor skills depends on growth and maturation of muscles, bones, and nerves. Teaching new skills is of little value until a state of readiness is reached.
10. Fine motor skills achieved during this stage are related to self-feeding, dressing, and playing. By end of this stage, children should be actively participating in dressing, washing, and brushing teeth.
11. Toilet training can be successful once the child has achieved a degree of maturity in sphincter muscles, nerves, and language.

12. Autonomy (independence) is a primary psychosocial task of toddlers. It encourages toddlers to make decisions, especially in their activities of daily living.
13. Two-year-olds need guidance and discipline. Caregivers should be consistent and repeatedly reinforce limitations. Limit setting should not deny children freedom but give them greater opportunity to explore.
14. Piaget suggests that toddlers interpret new experiences based on memory of previous happenings. This is referred to as preoperational thought.
15. Moral development depends on children imitating their parents' moral behavior and teachings.
16. Language acquisition develops along with memory and cognition. Young toddlers use one-word sentences. Three-year-olds have a vocabulary of approximately 1000 words and use multiple words in a sentence.
17. Toddlers need a well-balanced diet and good eating habits to support muscle and bone growth. Food amounts vary greatly depending on individual activity levels. Likes and dislikes are influenced by other family members' dietary habits. Most frequently accepted foods are those that the child can eat as he or she moves about.
18. Sleep needs decrease during this stage of development. However, short naps are still indicated. Bedtime rituals are common; they help reduce anxiety and give the child a sense of security. Nightmares should be handled in a consistent, comforting manner.
19. Play is the major means by which the child continues to explore and understand the world. Toddlers usually prefer parallel play. Imaginary play or playmates are common and normal at this stage.
20. Natural curiosity and the child's lack of ability to recognize danger make accident prevention of utmost importance at this stage. For this reason, toddlers require continuous supervision in all of their activities.
21. Regular physical examinations should be scheduled for toddlers. These visits should be at 18, 24, and 36 months. Dental exams are scheduled when the children have their complete set of deciduous teeth.

CRITICAL THINKING

Ms. Sterdowski brings Eric, 28 months of age, to the clinic because she believes that he is not thriving. She explains that he walks alone without assistance but only attempts to say about four words. His pronunciation is unclear to anyone outside of the immediate family members. She states that she has not been successful with potty training as of yet. She further explains that her older child had mastered all these skills and more by this age.

1. Describe the best approach to use to reassure Ms. Sterdowski that Eric is developing normally.
2. Based on expected growth timetables, what is the best interpretation of Eric's performance?
3. List three instructions that should be given to assist Ms. Sterdowski with accomplishing toilet training.

Multiple-Choice Questions

1. Toddlers' motivation to acquire and master most psychomotor skills is related to their need for:
   a. Balance
   b. Independence
   c. Sameness
   d. Dominance

2. Fine motor skills that should be mastered by 3 years of age include:
   a. Holding the spoon with the fist
   b. Using a crayon to draw a circle
   c. Drawing a complete face
   d. Recognizing dangerous situations

3. Toilet training depends on the child's ability to:
   a. Sit alone on the toilet
   b. Attain sphincter control
   c. Want to please the parents
   d. Properly digest a regular diet

4. The psychosocial task for the toddler according to Erikson is:
   a. Trust
   b. Initiative
   c. Autonomy
   d. Industry

5. Moral development in the toddler is based on:
   a. Innate instincts
   b. Mature behavior
   c. Promptly meeting needs
   d. Copying parental values

6. The type of play seen in the 2-year-old is:
   a. Solitary play
   b. Parallel play
   c. Cooperative play
   d. Competitive or team play

Suggested Readings

Fiese, BH: Playful relationships: A contextual analysis of mother-toddler interaction and symbolic play. Child Dev 61(3):1648–1656, 1990.

Goldson, E: The affective and cognitive sequelae of child maltreatment. Pediatr Clin North Am 38(6):1481–1496, 1991.

Greenspan, S: Clinical assessment of emotional milestones in infancy and early childhood. Pediatr Clin North Am 38(6):1371–1384, 1991.

Howard, BJ: Discipline in early childhood. Pediatr Clin North Am 38(6):1351–1369, 1991.

Ross, H, and Tesla, C: Maternal interventions in toddler-peer conflicts: The socialization of principles of justice. Developmental Psychology 26(6):994–1003, 1990.
Schuster, C, and Ashburn, S: The Process of Human Development: A Holistic Life Span Approach. JB Lippincott, Philadelphia, 1992.
Stipek, D, Gralinski, H, and Kopp, C: Self-concept development in the toddler years. Developmental Psychology 26(6):972–977, 1990.
Tamis-LeMonda, C, and Bornstein, M: Specificity in mother-toddler language-play relations across the second year. Developmental Psychology 30(2):283–292, 1994.
Tamis-LeMonda, C, and Bornstein, M: Individual variation correspondence, stability and change in mother and toddler play. Infant Behavior and Development 14:143–162, 1991.
Whaley, L, and Wong, D: Nursing Care of Infants and Children. Mosby-Year Book, St. Louis, 1991.

Chapter 6

Chapter Outline

Preschool

Key Words

adducted
conscience
cooperative/associative play
enuresis
initiative

Learning Objectives

At the end of this chapter, you should be able to:

- Describe the physical changes that commonly occur during the preschool years.
- List two gross motor skills characteristic of preschoolers.
- Describe the psychosocial task of the preschooler as outlined by Erikson.
- List the important guidelines useful in assessing a nursery school program.
- Describe the stage of cognitive development for the preschool child as presented by Piaget.
- List three appropriate snack foods for preschool children.
- Describe the type of play characteristic of the preschool years.
- List the safety concerns important to the preschool stage of development.
- Name two common behavioral concerns affecting preschoolers.

The preschool period generally refers to the period from 3 to 6 years of age. The rate of growth for the preschool period is best described as slow and steady. During this period preschoolers are focusing on refining their gross and fine motor skills, improving their vocabulary, and increasing their knowledge of their environment. Characteristically, children by this stage have mastered some autonomy and are moving toward a creative exploration of their potential. Preschool children are usually ready to spend more time away from their home and caregivers. Preschool or nursery school experiences often begin during this stage.

PHYSICAL CHARACTERISTICS

Height and Weight

The trunk and body lengthen, giving the child a taller appearance. On the average, preschool children gain 5 to 7 lb (2.3 to 3.2 kg) a year. Most children grow 2½ to 3 inches (6.75 to 7.5 cm) per year.

Body Proportions

During the preschool period there is a loss of some subcutaneous or adipose tissue, which accounts for the more slender look. Growth patterns vary with each child. The rate of growth for the extremities is faster than for the trunk, resulting in more adultlike proportions. The protuberant abdomen and exaggerated lumbar curvature (lordosis) disappear. The head and neck decrease in size in proportion to the size of the rest of the body.

Muscle and Bone Development

Rapid growth in the muscles accounts for approximately 75 percent of the weight gain during this period. Heredity, nutrition, and actual muscle use play a role in stimulating muscle growth and increasing muscle strength.

The hips gradually rotate inward, causing a more **adducted** foot position. The adducted movement causes the foot to move toward the center of the body. This results in a more erect posture and steady gait, making the child appear less awkward and clumsy.

During this stage of development, fat replaces the red marrow in the long bones. The marrow from this time on will be found in the flat bones of the body such as the skull, sternum or breastbone, vertebrae, and pelvic bones. Red marrow helps the body to produce blood cells.

Teeth

Deciduous teeth are important at this time and help prepare for the child's permanent teeth. This is a time when many children develop dental decay and a buildup of plaque. Care of the teeth should include daily brushing, flossing, and visits to the dentist at least every 6 months. Proper care and dietary practices may help to avoid excessive tooth destruction. Recommendations for promoting dental health are listed in Table 6–1. Preschool children should be encouraged to eat snacks that are low in carbohydrates such as apples, celery, carrots, and cheese. These snacks are nutritious and help prevent tooth decay.

Sensory Development

Visual acuity improves at 3 years of age to 20/20. The lack of depth perception contributes to some of the clumsiness that is still characteristic of the early preschool years. Depth perception and color detection are fully established by age 5. Maximum visual ability is achieved by the end of the preschool period.

Hearing matures at an earlier age. Children at this stage are better able to listen and to interpret and distinguish different sounds. Preschoolers, like toddlers, frequently develop ear infections. By the preschool years, the child may be capable of verbalizing and pinpointing ear discomfort. A simple ear examination followed by the prescribed course of antibiotics will treat an ear infection and prevent permanent hearing loss.

VITAL SIGNS

The preschool child's cardiovascular system enlarges to meet the general demands of the body. The average pulse rate for this period ranges from 90 to 100 beats per minute. The average blood pressure is 100/60 mm Hg. Hypertension may develop during the preschool period; therefore, blood pressure should be monitored during routine health assessments. The normal range for the respiratory rate is 22 to 25 breaths per minute at rest. This decrease in the rate of respirations is related to the growth in the lungs, permitting better efficiency.

DEVELOPMENTAL MILESTONES

Motor Development

Gross Motor Skills

Four-year-olds are capable of walking and running on their tiptoes. They can now hop and balance on one foot for 3 to 5 seconds and use alternating feet while descending stairs. Their development improves to the point where they can pedal a tricycle

TABLE 6–1
PROMOTING DENTAL HEALTH

Hygiene	Use small soft toothbrush. Use toothpaste that has fluoride. Brush daily in morning and before bedtime. Use back-and-forth motion while brush is against teeth and gum line.
Foods	Limit intake of high-sugar foods. Offer fresh fruits and vegetables daily.
Health supervision	Visit the dentist every 6 months.

quickly and navigate corners and turns. The 4-year-old likes to climb and jump from heights without demonstrating much fear, making constant supervision necessary (Fig. 6–1). A child of this age can catch a ball with extended arms and hands.

By 5 years of age the muscle coordination and strength increase, permitting the child to jump rope, skip on alternating feet, walk on a balance beam, and catch a ball with both hands. The preschooler's movements are now smoother and more efficient. This also enables the preschooler to begin certain sports, including soccer, skating, and baseball (Fig. 6–2). Five-year-old children are also capable of imitating and learning simple dance steps or other similar routines.

Fine Motor Skills

Four-year-old children are able to manage many of their self-care activities, including bathing, dressing, feeding, and toileting. These skills, which began in the toddler years, can now be performed with greater ease and dexterity. Although preschool children wash and dry their hands without supervision, they still require some assistance and supervision with the task of bathing.

Children at this age can manipulate their clothing, including buttons, zippers, and snaps. Children of 4 and 5 years old can readily recognize the front and back of their clothing. They still need help with tying shoelaces.

Preschool children can handle a spoon without inverting it, and many prefer to use a fork rather than a spoon. Many children are able to use a knife to spread butter or jelly. They still need help cutting up their food but like making their own sandwiches or pouring their own drinks. They are able to sit at the table for longer periods of time. Children of this age can learn about table manners. Preschool children usually can sit at meals for longer periods than toddlers. Whenever possible, families should eat together. Mealtime can be a time for family interaction, communication, and sharing of daily accomplishments.

At 4 years of age children are able to recognize their need to use the toilet but may require some assistance in manipulating their clothing and carrying out the necessary hygiene measures. By 5 years of age toileting has become a more independent

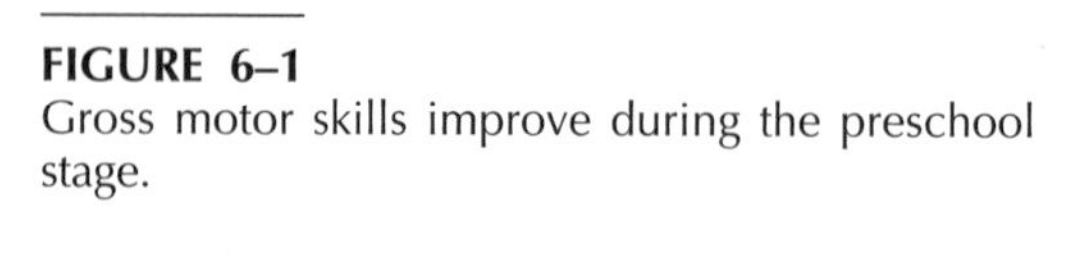

FIGURE 6–1
Gross motor skills improve during the preschool stage.

FIGURE 6–2
Preschoolers are now capable of learning more complicated sports.

practice. However, it is wise for parents to still supervise toileting to make certain that the child is washing his or her hands, wiping properly, and remembering to flush.

Fine motor development improves at 4 years of age to the point of allowing the child to draw a simple face and use a scissors to cut along a line. Five-year-olds can control their drawing to permit copying letters and printing their names (Fig. 6–3). Their drawings are more detailed and include not only a face but other body parts. The 5-year-old also has better control of scissors.

Sexual Development

During the preschool period children become aware of their genital organs and their sexual identity. As discussed in Chapter 3, children may become strongly attached to the parent of the opposite sex. They later identify with the parent of the same sex. Single-parent families should try to plan for the child to spend time with aunts, uncles, and other relatives or adult friends of the opposite sex. Children at this stage are also curious about the differences between the male and female bodies. They

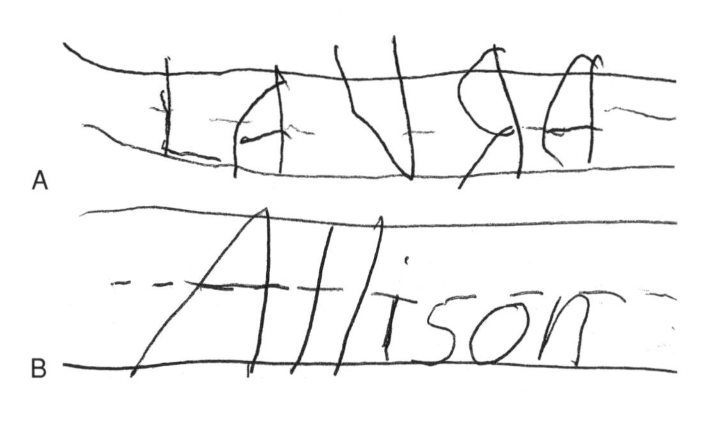

FIGURE 6–3
Progression of fine motor development from *(A)* 3 years to *(B)* 5 years.

may ask some questions about their bodies. Parents should respond simply and at the level of the child's understanding. For example, a preschool girl may ask why her baby brother looks different. Usually this age group is not looking for a detailed explanation. Parents should simply explain that boys look different. The child should be told that her brother has a penis. Using the correct terms and encouraging questions about sex helps keeps the lines of communication open between parents and children.

Masturbation is a common activity in both sexes and usually one of concern to parents. It is normal behavior, and caregivers must respond in a matter-of-fact manner to prevent instilling feelings of guilt in the child. As children mature and grow, they will develop mechanisms to channel sexual feelings, and this behavior will decrease.

Psychosocial Development

Initiative

According to Erikson, by preschool age the child has learned to trust those in the environment and has developed a sense of independence. Preschoolers pretend, explore, and try out new roles. Erikson referred to exploration as the task of **initiative**. Children enjoy playing and pretending to be many different people in their environment. It is important that young preschool children have good role models whom they may imitate and from whom they may learn. Teachers, health providers, clergy, and other adults in the community may serve as role models in addition to parents. Caregivers must allow children freedom to explore and try out roles they desire regardless of gender or stereotype. If not, the child can develop feelings of guilt; this may thwart their ability to grow and develop.

Discipline

Parents need to continue to set limits to protect children and property. Preschool children must not go into the street or hit others. As preschoolers develop a sense of initiative and guilt, they strive to follow rules and please parents. Discipline teaches children impulse control. If a child hits another child, the parents need to help the child learn to channel emotions into words. For example, a parent might say, "Stop it, you are very angry. Tell me why." This encourages the child to talk about his or her feelings.

Parents can also limit undesirable behavior by using positive reinforcement. In other words, parents can give a privilege or remove it, as the behavior warrants. Discipline should be of short duration because the child still has a limited concept of time. "Time out," as discussed in Chapter 5, can be useful in disciplining the preschooler.

Special Psychosocial Concerns

Jealousy. Jealousy, or sibling rivalry, discussed in Chapter 5, is a normal, inevitable behavior pattern seen at various stages of development. During the preschool stage children need affection, attention, and recognition. Unlike the toddler, the child of this age is better able to share and understand that he or she is not the only person in need of the caregivers' attention. Preschoolers who are involved with nursery school and other activities outside the home seem less threatened by a new baby or younger child at home. To help deal with sibling rivalry, parents should try to minimize it and to understand the child's feelings. Parents should make special time for the preschooler and give him or her a "special space" for toys and other meaningful items. Parents must always try to meet the preschooler's needs even if they are busy. Taking time out to tell the child a story or offer a special hug may be just enough to help the child adjust. The child of this age can also be given little "jobs" to help assist the parent with the care of other siblings. For example, the preschool child may be able to hold the bag, get the powder, or hand the parent what is needed. This helps strengthen the child's growing self-esteem.

Responses to divorce. Divorce is one of the common stresses affecting children during the preschool period. At this age, children may interpret the failing marriage as their fault. They usually have a strong wish to reunite their parents and may fantasize about this. Parents need to try to privately resolve their conflicts and spare the child undue emotional pain. Important measures include making the child feel loved and protected by both parties. The degree of emotional fallout from divorce can be minimized by having the noncustodial parent establish consistent and orderly visiting patterns. Children should feel that they have their own space in both parents' homes. Allowing the child to leave toys and clothing at the "other home" helps to reinforce a sense of belonging. See Box 6–1 for divorce hints for parents.

Preschool Education

The purpose of providing preschool education is to promote cognitive, motor, and social development. Preschool provides a place where children can form friendships

BOX 6–1

Divorce Hints for Parents

1. Only a firm final decision should be presented to the children.
2. Avoid casting blame or criticizing the other partner.
3. Avoid involving the children in any divorce-related matters (financial, legal, and so forth).
4. Continue routines as much as possible.
5. Reassure the children that they are not to blame for the divorce.
6. Offer love and support to the children.
7. Encourage open expression of feelings.
8. Create a space for the children's belongings in the noncustodial parent's home.
9. Avoid interrogating the children after visits with the other parent.
10. Seek supportive services for both parent and children, as needed.

and begin to learn to get along with peers. Starting school is a new experience for young children (Fig. 6–4). The first day should be brief and in the company of the parent until the child appears to be acclimated to the surroundings. For some children this is especially trying and requires special understanding and patience. No activity should be forced on a child. If a child is not ready for preschool, socialization such as forming friendships and getting along with peers is unlikely to occur. Parents and school staff can work together to assess the readiness of the child. Signs of preschool readiness include mastery of toilet training, ability to tolerate brief periods of separation from parents, and increased communication skills. Children eventually gain confidence in themselves and become interested in participating with peers (Fig. 6–5). To better prepare the child for preschool, parents can begin by familiarizing the child with the school location, visiting the school before the start, and offering ample opportunity to discuss their feelings.

Parents should consider several factors when selecting a preschool for a child. First, the parents should consider the location, cost, and schedule to make certain that the school fulfills these needs. Parents should determine what philosophy forms the basis of the preschool program.

It is recommended that the program be student-driven instead of teacher-driven. This means that the curriculum should be designed to meet the needs of the children first. This design places the children in the center of activities. The philosophy must appreciate the physical, cultural, cognitive, and emotional differences among the children. To best indicate the school's appreciation for these differences, the school should recognize and discuss holidays, introduce foods from different cultures, and value and celebrate diversity. The National Association for the Education of Young Children (NAEYC) recommends that at least one teacher hold a degree in early childhood education. NAEYC recommends a teacher-child ratio of one adult for every four to six 2-year-olds, with a maximum group size of 12, or one adult for every seven to ten 3- to 4-year-olds, with a maximum group size of 20.

Parents should make at least one unannounced visit to the preschool and be free to visit at any time. In addition, the school should have adequate play space to allow

FIGURE 6–4
Preschoolers show the ability and readiness to learn.

FIGURE 6–5
Preschoolers like being with peers.

freedom of movement. The supplies at the preschool should include enough blocks of all sizes as well as other hands-on materials such as clay, sand, wood, water, and puzzles. The entire classroom area should be clean but not an environment that is prohibitive to exploration and play. Bathroom facilities must be close by and accessible for preschoolers. See Box 6–2 for a preschool safety checklist.

Cognitive Development

According to Piaget, the preoperational stage of development begins in the toddler period and extends through the school-age years. It is limited in several ways. First,

BOX 6–2

Preschool Safety Checklist

1. Playgrounds should be secured with a fence.
2. Equipment must be free of sharp edges, in a good state of repair, and routinely inspected.
3. Fire alarms and fire extinguishers must be present and working.
4. Emergency exits and evacuation plans must be clearly posted.
5. Toys must be age-appropriate and in good condition.
6. Staff should be trained in first aid and CPR.
7. Bathrooms must be kept clean and easily accessible for young children.
8. Water temperature in faucets must be no higher than 110°F to prevent accidental burns.
9. Pick-up and drop-off sites should be secured for safety.
10. Children must be observed to leave only with authorized caregivers.

the preschooler is unable to focus on several aspects of a stimulus. For example, if someone is wearing a mask, preschool children don't recognize the person and become frightened. Piaget called this centration and described it as the child focusing or centering attention on only one cue. Piaget further believed that preschoolers lack reversibility, or the understanding of how two actions may be related to each other. For example, preschoolers may watch liquid being poured from a short fat beaker into a tall thin beaker, but still maintain that the tall beaker contains more fluid. Preschoolers are unable to understand that the physical attributes of an item remain the same despite superficial changes in its appearance. This is easily demonstrated by asking the preschooler to tell you if the amount of Play-Doh has changed when you simply change its shape.

At this stage the child continues to develop language and memory. The preschooler is still somewhat egocentric but is now able to share, take turns, and follow rules.

Preschoolers can form general concepts but cannot reason formally. Their reasoning appears to be based on their earlier experiences. For example, preschoolers, like toddlers, become concerned when they see their mother dressing up in her "going-out" clothes, knowing that this means that she is leaving. If trust and autonomy have developed sufficiently, however, the preschooler is able to tolerate separation and understand that Mommy will come back. Limited reasoning is also evident in the preschooler's concept and understanding of time.

During the preschool years children often pretend and are highly creative (Fig. 6–6). When the 4- or 5-year-old tells a story, parts of the story may be embellished and enhanced by the child's overactive imagination. The thinking of these children may also be magical and give them the illusion that they are all-powerful. Sometimes they develop feelings of guilt over their "bad" thoughts, which they believe

FIGURE 6–6
Preschoolers are highly creative and like to do arts-and-crafts projects.

may have caused an accident or other happening. It is also common for preschoolers to believe that if they become ill or injured, it is some type of punishment for their wrongdoing. Parents and healthcare workers need to explore preschool children's feelings to minimize their guilt and misunderstanding.

Preschoolers have a longer attention span than toddlers. This permits the preschool child to stay at an activity for longer periods. Toddlers may have difficulty sitting still and listening to an entire story, whereas preschoolers listen attentively, often memorize the story, and don't let you skip even a word.

Moral Development

According to Kohlberg, the preconventional stage of moral development begins at about age 4 years and continues to about age 10 (see Chap. 3). Moral development is a process in which children learn by modeling and imitating adult behaviors. For this reason it is imperative that parents set good examples and have consistent patterns of interaction.

According to Freud, children at this stage begin to develop a superego or **conscience**. The conscience gives them the capacity for self-evaluation and criticism. The superego becomes the moral dimension of the child's personality. It is at this stage that children begin to emulate beliefs, values, and ideals based on what is learned in their home and immediate environment. Caregivers must help children understand the cause and effect that certain behaviors may have on others. Preschool children should be reminded that if they grab a toy away from a playmate, the action will cause an unfavorable response.

The vivid imagination of preschoolers can make it difficult for them to distinguish fantasy from reality. They often tell stories that have threads of truth woven together with untruths. Preschoolers are just beginning to recognize that deliberate "lying" is a bad thing to do.

Communication

Language and speech become more sophisticated during the preschool period. Whereas toddlers make their needs know by gestures and the use of a few words, preschoolers can use nouns, verbs, and adjectives in their sentences. At 4 years of age children can form sentences using three to four words. Most 5-year-olds can form sentences containing five or more words. The 5-year-old's vocabulary contains between 2000 and 2400 words.

Preschool children may exhibit some difficulty with the pronunciation of certain words. This is to be expected and should be treated by the adult by gently correcting the child's mispronunciation without criticism. Putting undue stress on the child's ability to speak clearly may lead to stuttering and hesitancy. Some hesitancy is a normal pattern of speech development.

Children of 3, 4, and 5 years are usually very talkative. They like monopolizing the conversation, and will even talk if no one is listening or answers them. Preschoolers question even if they know the answer. They can express past, present, and future with an improved understanding of time. Because of this they can be put off until later better than the toddler. For example, if preschoolers riding in a car ask for a drink, they can be told that in just a few minutes they will be able to have one. Unlike the toddler, the preschooler can understand basic time concepts and delay their gratification. They also enjoy talking on the telephone. At 3 years of age their phone skills are limited to talking without interactive conversation. In other words, the 3-year-old will tell of a happening but not be able to answer questions asked over the phone. By 5 years of age the child can converse over the telephone.

Preschool children can be taught their full name, address, and telephone number. This is the time for parents and teachers to instruct children on how to respond to emergencies. This age group can be taught how to dial 911 to initiate the help system if needed.

Preschoolers learn by imitating others in their environment. Their vocabulary increases by repetition and practice. It is important that adult role models use appropriate words to instill positive influences on the growing child. As preschoolers look for the right word, a hesitation or stuttering may occur. This usually disappears within a few months. Parents should be patient when this occurs and listen without hurrying or labeling the child a stutterer.

When the child repeats an unacceptable or bad word, the parent needs to simply correct the child without making a fuss over this behavior. Drawing attention to the "bad" words serves only to reinforce the child's negative behavior.

Helpful Hints: Strategies to Stimulate Language Development

- Read to your child.
- Encourage storytelling.
- Play naming games.
- Gently correct mispronunciations.

NUTRITION

The nutritional requirements for the preschool years are similar to the requirements for the toddler years. The child will need all of the basic nutrients outlined in the food pyramid. The average caloric needs for this age are approximately 1800 calories per day, divided over the course of the day. Like toddlers, most preschool children do best when they have three meals and three snacks daily.

Certain issues may be a concern during this stage, including the continuation of food fads carried over from the toddler years, rebellious behavior, and periods of diminished appetite. Many parents become concerned about the amount of food consumed by their child. Parents should not expect the preschooler to eat an adult-size portion of food. Some children develop strong food preferences that may limit the type of foods that they will eat. By age 5 many children begin to develop food habits similar to those of their peers. Fast foods and highly advertised foods are known favorites for this age group. Parents must exercise caution when planning meals in fast-food restaurants as many of these food choices are very high in calories and fat.

The best diet for young children includes food sources containing proteins, carbohydrates, vitamins and minerals, and limited fats. Milk is still an important food because of its calcium content. Preschoolers need at least 1 pint of milk per day to meet their daily calcium requirements. It is also important that snack foods be selected for their nutritional value as well as appeal. Good food choices for snacks may include the following: fresh or dried fruits, vegetables in bite-size pieces, cheese, or yogurt. As stated earlier, it is best not to offer as a snack candy or other excessively sweet foods. These foods are not good for the health of the teeth and have a tendency to spoil the child's appetite for regular meals. See Table 6–2 for a typical preschool diet.

Mealtimes should be pleasant and allow family interaction. If the mealtime is very long, the preschooler may not be able to sit through it without getting restless and fidgety.

Helpful Hints: Fostering Good Eating Habits

- Set good examples.
- Never force eating.
- Don't use food as a bribe or reward.
- Eat with the child.
- Provide a relaxed atmosphere.
- Encourage child's help in preparation and cleanup.
- Provide positive reinforcement.
- Allow child to eat foods in any order.

TABLE 6–2
SAMPLE PRESCHOOL DIET

Type of Food	Amount per Day
Dairy foods	
Milk	4 oz
Cheese	1/2–3/4 oz
Yogurt	1/4–1/2 cup
Protein foods	
Meat, fish, or poultry	Two servings, 1–2 oz
Eggs	3 eggs per week
Peanut butter	1–2 tablespoons
Legumes, dried peas, beans, cooked	1/4–1/2 cup
Vegetables and fruits	
Vegetables, cooked	Four servings, 2–4 tablespoons; include 1 green or yellow vegetable
Vegetables, raw	Several pieces
Fruits, canned	4–8 tablespoons
Fruits, raw	Several small pieces
Grains	
Bread, enriched white, whole wheat	1–2 slices
Multigrain crackers	3–4
Cereal, cooked or dry	1/2 cup
Pasta, rice	1/4–1/2 cup
Fats	
Bacon	1–2 slices
Butter	1 teaspoon

SLEEP AND REST

The average preschool child requires about 10 to 12 hours of sleep each night. Children in this age group are usually very active and busily engaged in some activity or game. Many are too busy and active to have time for a nap even though they still need to rest and replenish their energy. Some children now take only one nap per day, and many don't do this on a daily basis. The average nap at this age lasts 30 to 60 minutes. By evening the child is generally overtired and very much in need of an early bedtime. Many children during this stage have difficulty falling asleep. Following a consistent bedtime routine helps minimize the conflict or debate over bedtime practices. Each child should be expected to follow the bedtime ritual regardless of the day of the week or other distraction. It may also be necessary to limit the ritual because some youngsters lengthen the ritual as a means of putting off the inevitable.

Preschool children often wake up during the night with frightening nightmares. Parents should gently reassure the child that he or she is safe and not alone. After sitting with the child until he or she is relaxed, the parent should then proceed with the bedtime routine and encourage the child to go back to sleep. Some children experience a more extreme form of nightmare called night terrors. When this occurs, the child suddenly sits up screaming, not fully awake. The child's appearance may be very frightening to the parents. This may be accompanied by a loud scream, rapid breathing, rapid heart rate, and profuse perspiration. The child may be inconsolable for a brief period. These episodes are usually followed by the child relaxing and falling back into a quiet sleep. The child is unable to recall the event in the morning. Night terrors are not thought to represent any emotional stress and disappear without intervention.

PLAY

Preschool children engage in **cooperative** or **associative play**, which requires that the child be able to understand limited rules. By this age children usually have developed some social skills that permit them to begin to share and take turns. They have mastered basic communication skills, which further enable them to express their desires. They enjoy being with peers and interacting with them during play.

When observing preschool children at play, one can clearly see different personality traits emerge. For example, some children are more dominant and others are passive and follow another child's lead. Some children are more cautious and timid when trying out new activities, whereas others aggressively attack any new activity without regard for danger. Still others may not interact well with peers or may even be excluded from peer activity. Unless the child is victimized or appears unhappy, he or she should be allowed to socialize at his or her own pace. Caregivers should understand that children have different personalities and different social needs. For example, some children need to be in the center of all activities, whereas some prefer to be observers or to play alone.

To help further develop gross motor skills such as jumping, running, and climbing, the preschool child uses toys such as jungle gyms, tricycles, and Big Wheels. It is now possible to teach the preschooler swimming, skating, soccer, and other sporting activities. It is important, though, that parents not force their own competitive needs on their children. Just as growth patterns vary with each child, so do gross motor skills and innate abilities. Some children are more coordinated and skillful than others. Fine motor skills are enhanced through the use of many different playthings. Puzzles, construction sets, and computer programs help children at this age learn to manipulate and coordinate their small motor movements. Computer programs and other interactive games also help stimulate the preschooler's thinking skills, reasoning, and memory. The preschool years are a time for building confidence. Children at this stage frequently say, "Look at my tall tower" or "Listen to me sing the ABCs." "Watch me" is a common demand. At this age they seek approval and recognition from adults.

Child-size kitchens and tool corners permit children to use their imagination, try on roles, and pretend to be grown up. Regardless of the type of plaything used, it is important that children be taught to clean up their play area and take care of their toys. This helps to teach the child responsibility and initiative, which can later be transferred to other situations.

Preschoolers still use symbolic play as described in Chapter 5. A great amount of time spent at play is in activities that use their imagination. Dramatic play may involve dressing up in different clothes, exploring different roles, and imitating adults (Fig. 6–7). Common activities may also include playing with dolls or tools.

Television is sometimes used as a time filler and for entertainment. Parents should screen the types of programs that their children view. Many programs have educational value, whereas others may have little or none. Educational programs might also foster good social relationships that are imaginative and help enhance the child's imagination. Historically, television shows involving puppets have been useful in stimulating the imagination and creativity of preschool children.

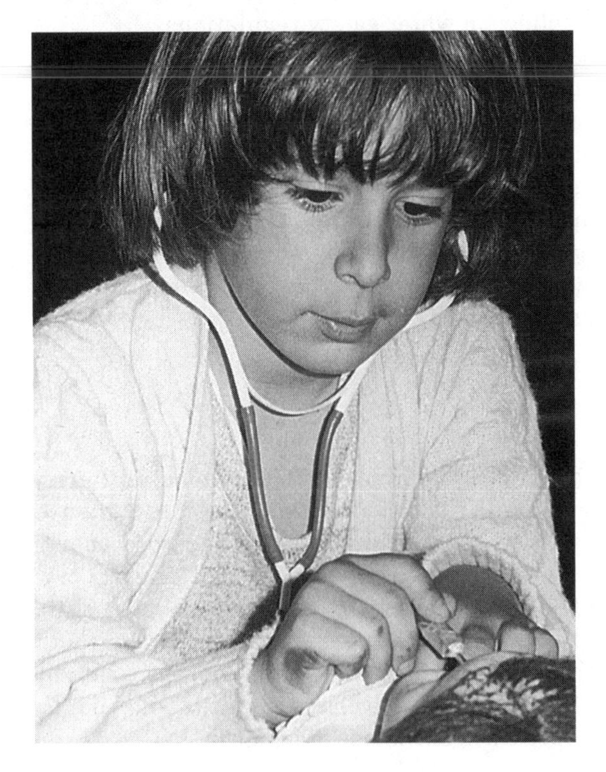

FIGURE 6–7
Preschoolers like to try different roles and use their imagination.

SAFETY

The better-coordinated preschool child is less likely to fall than the awkward toddler. This age group is also more aware of certain dangers and limitations. Most of them can recite a long list of "Nos" or things they cannot touch or do. Even so, they still need adult supervision and continuous reminders about potential environmental hazards. The safety precautions outlined in Chapter 5 for the toddler apply to the preschool child as well.

Clothing items must be carefully inspected to make certain that they are safe for the active child. Clothing must allow freedom of movement and be nonrestrictive. Recently, several deaths have occurred as a result of clothing strings, belts, or loops catching on playground items and resulting in strangulation deaths. It is the responsibility of both the clothing manufacturers and caregivers to help look for these potential dangers to prevent accidents. Federal regulations now require that children's nightwear be flame-resistant as an additional means of providing for child safety in the home.

Motor vehicle accidents are still the major cause of accidental death in preschool children. Preschool children must always be restrained in proper car safety seats. Depending on the specific seat and size of the child, some will be ready to move to the adult safety restraint. It is best to check with the car manufacturer for guidelines about use. Because preschoolers like to imitate adult behaviors, adults must always use their own car safety restraints.

As the child becomes more active and involved in various games and activities, there is an increase in the number of sports and recreational injuries. Preschool children are less likely to drown in the bathtub but more likely to get into danger outdoors near pools or other bodies of water. Preschool age children should be given swimming lessons and instructions to follow on water safety regulations.

Parents must begin to educate their children about certain dangers, such as talking to strangers and accepting candy, rides, or money from strangers. It is also necessary to teach children about their "private body parts," including the rule that no one should be allowed to touch them. The environment in the home must permit the child access and freedom to discuss their concerns and worries without shame or ridicule. Many communities have enacted safety programs that fingerprint and photo-ID children as a preventive measure in case of child abduction or for help in searching for a missing person. Parents may contact their local police departments for information or help in establishing these types of programs.

HEALTH PROMOTION

By the preschool age, the child needs to be given booster shots for diphtheria, pertussis, and tetanus (DPT) and trivalent oral polio vaccine (TOPV) to ensure immunity and protection against these diseases. Children at this age need yearly preventive healthcare visits to supervise their physical and social growth and development. See Box 6–3 for a sample of preschool health screening assessments. Because preschoolers have increased exposure to other children, they are more likely to spread simple infections such as the common cold. They should be taught common hygiene principles such as handwashing, covering the nose and mouth when sneezing or coughing, and the importance of the use of tissues.

BOX 6–3

Preschool Yearly Health Screening

A complete examination should include the following:

- Physical examination
- Health history
- Physical, nutritional, and psychosocial assessment
- Vision and hearing testing
- Cardiac screening
- Blood screening
- Urine testing
- Immunizations
- Tuberculosis screening
- Dental supervision

Common Preschool Concerns

Thumb Sucking

Thumb sucking is thought to be a primitive, instinctive behavior that may fulfill the child's sucking and comfort needs. One concern with prolonged thumb sucking is that it may cause malalignment of the teeth. For this reason, many parents try to discourage this habit, often unsuccessfully. Sometimes other comfort objects such as a teddy bear, soft doll, or blanket can replace the thumb and still provide the child with a sense of security. Usually by school age this behavior lessens and eventually disappears.

Bed-wetting

Bed-wetting, also known as **enuresis**, is a problem that is seen more often in boys than in girls. Enuresis cannot be even considered until after toilet training is well established. The cause of enuresis is not fully understood. Stress and illness in the child appear to make it worse. If this behavior persists, it is important that the child be given a complete medical examination to help rule out any underlying pathology. The best approach in dealing with the problem of bed-wetting is one that minimizes each episode and avoids making the child feel guilty or ashamed. Parents can sometimes help prevent these accidents by taking the child to the toilet in the evening hours and limiting the child's fluid intake after 5 PM. Punishment or ridicule lowers the child's self-esteem and morale and should be avoided.

Fears

Preschool children frequently have fears of the dark, mutilation, and abandonment. All childhood fears should be approached in a similiar manner. First, caregivers must acknowledge the child's fears. Reassurance and reality reinforcement are essential in helping children cope with their fears. Some preschoolers experience fear of the dark. This may cause bedtime hassles or interrupt sleep. A simple night light left on in the child's room or in a hallway may lessen the discomfort.

Fear of mutilation becomes evident at the time of injury or during hospitalization. Bleeding from a small scrape seems particularly frightening to this age group. Band-Aids can help cover the site of the injury, making the child feel better.

Fear of abandonment occurs at this age. Preschoolers may get hysterical if a parent is a few minutes late in picking them up at school. They respond as though the parent is never coming back. However, they show increased independence by thinking nothing of walking away from their parents in a store or on the playground. They think it is all right if they wander out of sight, but their response is different if the parent walks away.

SUMMARY

1. The rate of growth for the preschool period is best described as slow and steady.
2. Children grow 2½ inches per year. The average weight gain is 5 to 7 lb a year. The trunk and body lengthen, giving the child a taller appearance and more erect posture.
3. Care of teeth during the preschool stage is important to promote healthy teeth in future stages.
4. Visual acuity improves and hearing matures during this stage.
5. The structural makeup of the child's ear continues to account for the high incidence of middle-ear infections in this age group.
6. Normal pulse rate is between 90 and 100 beats per minute and blood pressure is about 100/60 mm Hg. The normal respiratory rate is 22 to 25 breaths per minute at rest.
7. The focus during the preschool years is on motor skills, improving vocabulary, and increasing knowledge about the environment.
8. Four-year-olds are capable of walking and running on their tiptoes, hopping, and balancing on one foot. Preschoolers can pedal a tricycle and like to climb and jump. By age 5, they can skip, walk on a balance beam, and catch a ball.
9. Improved fine motor development allows 4-year-old children to accomplish self-care.
10. Preschoolers find pleasure in examining and exploring their bodies. They are now very curious about the differences between male and female bodies.
11. Masturbation is common for both sexes during this stage. Parents should respond in a matter-of-fact manner and not instill guilt feelings in their child.
12. The psychosocial task of the preschool period is the development of initiative.
13. Preschool children need discipline in order to learn impulse control.
14. Preschools should be selected on the basis of their philosophy, location, and cost.
15. Jealousy is a normal pattern of behavior seen at various stages of development.

16. Divorce is one of the common stresses that affect children. Preschool children often blame themselves and have a strong wish to reunite the parents.
17. Cognitive development is at the preoperational stage. At this stage the child continues to develop language and memory. The thinking style of the preschool stage is often described as magical, giving the child the feeling that he or she is all-powerful.
18. Preschool children are developing their conscience and have a beginning capacity for moral reasoning. Moral reasoning is learned mainly by imitating parents and other adults. It is therefore important that the adult role model use appropriate words and set good examples for the child.
19. Communication is more sophisticated during this period of development. Children are very talkative and can be taught their full name and address and how to respond to emergencies.
20. The average caloric need for this age is 1800 calories a day, divided over the course of the day. By age 5 many children have developed food habits similar to those of their peers. The diet for this age group should include food sources containing proteins, carbohydrates, vitamins, minerals, and limited fats.
21. Children at this stage are very active and require an average of 10 to 12 hours of sleep each night. All preschool children need consistency with their night rituals. Nightmares and night terrors are common during this stage of development.
22. Preschool play style is known as cooperative or associative. Children are able to share, take turns, and follow simple rules.
23. Toys should be selected to help stimulate fine and gross motor development. Just as growth patterns vary with each child, so do motor skills and innate abilities, making some children more coordinated than others.
24. Safety continues to be a major concern for preschool children, and they still need constant reminders about potential environmental hazards and adult supervision.
25. During the preschool years, children need to be given boosters for DPT and TOPV to help ensure immunity and protection against these diseases. Children at this stage need yearly preventive healthcare visits and screening to supervise their physical, emotional, and social development.
26. Thumb sucking is thought to be a primitive and instinctive behavior. If it becomes a prolonged habit, it may cause malalignment of the child's teeth.
27. Bed-wetting, also known as enuresis, is a problem seen more often in boys than in girls. Stress and illness in the child seem to make it worse.

CRITICAL THINKING

Mrs. Hyatt attends a local community support group supervised by the pediatric nurse practitioner for new and working mothers. Mrs. Hyatt verbalizes her concern about her 4½-year-old son. He appears to be slow in mastering language. He is presently enrolled in a preschool program. Although his speech has improved since he began school, he demonstrates marked stuttering. Mrs. Hyatt also reports that her son repeatedly uses bad language. She is very upset and admits that she doesn't know how to handle this.

Mrs. Hyatt assures the nurse that neither she nor her husband uses this type of language in the home.

1. What reason would you give Mrs. Hyatt for her son's:
   a. Bad language?
   b. Stuttering?
2. How would you instruct Mrs. Hyatt to respond to her son when he:
   a. Uses bad language?
   b. Stutters?

Multiple-Choice Questions

1. The erect posture and steady gait seen in preschool children may be due to:
   a. Movement of the foot away from the center of the body
   b. Exaggerated lumbar curvature of the spine
   c. Movement of the foot toward the center of the body
   d. Increased fusion of the spinal bones

2. Ear infections are more commonly seen in children because of:
   a. Their increased exposure to infection
   b. Their smaller earlobes, proportionate to those of adults
   c. Their shorter eustachian tube
   d. A decrease in the number of white blood cells

3. The type of play seen at the preschool age is:
   a. Parallel play
   b. Associative play
   c. Solitary play
   d. Isolated play

4. Occasional periods of masturbation in a 4-year-old suggest:
   a. Pathology in the child's personality
   b. A normal pattern of development
   c. History of sexual abuse
   d. Impaired cognitive development

5. The average daily caloric intake recommended during the preschool years is:
   a. 1000 calories
   b. 2500 calories

c. 1800 calories
d. 500 calories

6. Nightmares and night terrors differ in the following way:
   a. With nightmares the child remains inconsolable for a long period.
   b. With night terrors the child remembers the event in detail.
   c. Nightmares are usually accompanied by rapid respirations and rapid heart rate. With night terrors the child has no recall of the event.

Suggested Readings

Delaney, S: Divorce mediation and children's adjustment to parental divorce. Pediatric Nursing 21(5):434–437, 1995.

Heath, C: The key question to ask when choosing a preschool. Parents 70(11):118–120, 1995.

Ignoe, JB, and Giordano, BP: Health promotion and disease prevention secret of success. Pediatric Nursing 18(1):61–62, 1992.

Jones, NE: Prevention of childhood injuries. Part I. Motor vehicle injuries. Pediatric Nursing 18(4):380–382, 1992.

Keltner, BR: Family influences on child health status. Pediatric Nursing 18(2):128–131, 1992.

Kutner, L: Non shaming techniques to end bed-wetting. Parents 70(12):93–94, 1995.

Kutner, L: Let's make a deal. Parents 71(1):70–71, 1996.

Pritchard, P: Behaviorial work with pre-school children in the community. Health Visit 67(2):54–56, 1994.

Standford, DA, Ahlrichs, J, Carmicle, C, and Wells, PW: Extended day program bringing preschool to the hospital. Pediatric Nursing 19 (3):238–241, 1993.

Wong, DL: Whaley and Wong's Essentials of Pediatric Nursing. Mosby-Year Book, St. Louis, 1993.

Chapter 7

Chapter Outline

School Age

Key Words

epiphyseal cartilage
industry
latency
malocclusion
ossification
puberty
reciprocity
saturated fats
school phobia
scoliosis
somatic
team play

Learning Objectives

At the end of this chapter, you should be able to:

- List four physical characteristics common to school-age children.
- Describe three developmental milestones common to school-age children.
- Describe the psychosocial task identified by Erikson for the school-age period.
- Describe the cognitive levels of functioning during the school-age period.
- Describe moral development in school-age children.
- List three factors that help contribute to the health of school-age children.

The period of development known as *school age*, the middle years, or *late childhood* starts with the child's entry into formal education and ends with the onset of puberty, roughly from ages 6 to 11. **Puberty** commonly refers to the developmental period in which the body prepares for the changes necessary for reproduction. Five significant accomplishments occur during the school-age period:

- Growth becomes slow and steady.
- Children move away from the family toward peer relationships.
- Children become less self-centered and more goal-directed.
- Deciduous teeth are lost and the permanent teeth appear.
- Sexual tranquility replaces sexual curiosity and preoccupation.

PHYSICAL CHARACTERISTICS

Height and Weight

The school-age period begins with slow, consistent growth and ends with a growth spurt just at the time of puberty. The average expected growth rate for the child during this period of development is 2 to 3 in (5 to 7 cm) per year. Weight increases on an average of 4.5 to 6.5 lb (2 to 3 kg) per year. The average 6-year-old girl measures 45 in (116 cm) and weighs 46 lb (21 kg), while the average 10-year-old girl measures 59 in (150 cm) and weighs 88 lb (40 kg). Boys may appear taller and heavier in the early school-age period, but for a brief time toward the end of this stage, girls are taller and heavier than boys.

Bone and Muscle Development

Bone growth and maturation can be affected by several factors, including gender, race, nutrition, and general state of health. In girls, the bones mature 2 years earlier than in boys. Black children, in general, show earlier bone development than do white children. Growth in the long bones stretches the ligaments and muscles, causing most children to experience "growing pains," mostly at night. The child's arms and legs lengthen, producing a thin, spindly appearance.

School-age children's posture changes because their center of gravity shifts downward as their muscle strength increases. The abdominal muscles also grow stronger, causing the pelvis to tip backward. The chest broadens and flattens, while the shoulders continue to appear rounded. Exercise encourages muscle development and improves strength and flexibility. Poor posture causes fatigue and may indicate minor skeletal pathology. Health screening for defects in the skeletal system is discussed under the section entitled "Health Promotion."

Generally, muscle mass and muscle strength increase, but the school-age child's muscles are still relatively immature and easily injured. Fine and gross motor skills show marked improvement, allowing the child to be more independent and self-sufficient both in the home and at school.

Sensory Development

Visual maturity is usually achieved by 6 to 7 years of age. Peripheral vision and depth perception improve, permitting better hand-eye coordination. School-age children no longer require large print in books and schoolwork.

Dentition

An important hallmark of this stage is the loss of the deciduous teeth and the appearance of the permanent teeth (Fig. 7–1). Children should be told in advance that they will lose their baby teeth so that they are not frightened when it happens. This can be done when the first tooth starts to loosen, usually at 6 to 7 years of age. Parents can emphasize that this is a sign that the child is growing up. Some parents play the "tooth fairy game," rewarding the child with money for each lost tooth. The first tooth to fall out is the lower central incisor. Teeth should not be pulled or forced to fall out but allowed to progress naturally.

The permanent teeth appear in the same order as the deciduous teeth. They usually appear very large in relation to the rest of the facial structures. The result is what is sometimes called the "ugly duckling" stage of development. Up to 75 percent of children have some degree of **malocclusion**, or malposition of the teeth. This may affect their chewing, facial relaxation, and appearance. Children should visit the dentist regularly, at least every 6 months, to have their teeth inspected and cleaned and any dental disease corrected. Daily dental care should include regular toothbrushing after meals and before bedtime. The use of fluoride toothpaste is strongly recommended to help decrease the incidence of dental caries. Dental caries begins with the buildup of plaque around the tooth surface and margins. Regular brushing and limiting children's intake of concentrated sweets help prevent plaque

FIGURE 7–1
The school-age period is marked by the loss of the deciduous teeth.

formation. Certain snack foods such as apples, raw carrots, and sugarless gum can help reduce plaque formation.

Development of the Gastrointestinal and Nervous Systems

Because the gastrointestinal system matures during this stage, school-age children have fewer digestive intolerances and disturbances than younger children. As the capacity of the stomach increases, the child needs to eat less often. Three meals a day is now sufficient.

The nervous system continues to mature, as evidenced by the child's improved motor skills and expanded cognitive processes. The senses of taste, smell, and touch fully mature, making the school-age child more discriminatory. At this stage children develop many distinct food preferences based on personal taste and peer influences.

Development of the Immune System

The school-age period is marked by the maturing of the immune system, producing a peak in the child's antibody levels. There is an increase in the amount of lymphatic tissue in the nasopharynx—the tonsils and adenoids. These tissues may be disproportionately large, but unless they are causing infection or obstruction, surgical removal is not recommended. When children begin school, they are exposed to a greater number of microorganisms, and as a result they often have an increased incidence of upper respiratory tract infections. Once their immune systems adjust to the increased exposure, their resistance improves. The school-age period is generally a healthy period of development.

VITAL SIGNS

Because school-age children's hearts are small in proportion to their body mass, they may feel tired after strenuous exercise. During this stage the heart rate decreases to an average of 90 beats per minute. Functional (innocent) heart murmurs may be present in 50 percent of school-age children. These murmurs do not usually require intervention. Blood pressure readings are generally higher than at the earlier stages—usually 100/60 mm Hg—because of the development of the left ventricle. Hemoglobin and hematocrit levels also increase slightly, whereas red and white blood cell counts decrease slightly during this period of development.

The respiratory system of the school-age child continues to develop; by 8 years of age the child's alveoli (air sacs) are fully mature. The normal respiratory rate decreases; the average resting respiratory rate is 20 breaths per minute.

DEVELOPMENTAL MILESTONES

Motor Development

During the school-age period there is a marked increase in muscle mass and muscle strength and significant improvement in gross and fine motor skills. School-age children can run faster, farther, and for longer periods. At this stage they are able to jump higher and throw farther and with more accuracy than children in younger age groups. Most children are stronger and better coordinated at this stage. Gender differences exist in motor skills. On the average, boys are stronger and better at running, jumping, and throwing, and have greater endurance than girls. Girls are better at balance and coordination than boys. Girls perfect their fine motor skills before boys perfect theirs. Motor accomplishments are very important to both girls and boys during this stage.

School-age children have developed enough proficiency in gross and fine motor skills to permit independence in many areas, including play, self-care, school, and the home. Although 6-year-old children appear more grown-up and independent, they are easily frustrated and fatigued. It is not uncommon for them to cry and become very irritable and infantile. Children of this age group often play on their own and select activities that they find enjoyable. Many of their new-found skills can be accomplished without parental assistance. Some of the skills that validate their independence include swimming, skating, and bicycle riding. During this stage children show competence in performing necessary self-care activities such as bathing, dressing, and feeding themselves. School-age children learn to write, draw, dance, and develop many other creative talents.

Most 6- to 7-year-olds can print letters and their name, as well as throw, catch, swim, and run with better control. Children of 6 and 7 years old continue to learn to tie their shoelaces.

The gross motor skills of 7- and 8-year-olds improve, permitting smoother movements while running, jumping, and skipping. Many at this stage are capable of using a ball and bat with greater control and accuracy.

Fine motor skills continue to improve, and by 8 years of age fine movements become steadier and more controlled. Children now prefer a pencil or pen to a crayon and can print smaller and learn script lettering. With their improved fine motor development, they can now begin to learn to play many musical instruments.

Children of 8 to 9 years are usually outgoing, talkative, and enthusiastic. They are ready to take on any project regardless of their capability. This fearlessness may put them at greater risk of injury. At this stage they appear more graceful and have smoother motor coordination (Fig. 7–2). Their strength and endurance also increase, improving their motor performances. Eight-year-olds practice a skill longer and with more of a sense of commitment. Once they master the skill, they are ready to show off their talents. Various new activities may be attempted, such as gymnastics, karate, ballet, and other dance forms. Their fine motor skills also improve to now permit mastery in the areas attempted at an earlier age. There no longer seems to be a random selection of games and activities. Instead children select activities based on their specific interests and likes. By the end of this stage, their physical strength is almost equal to that of an adult. Endurance and skills improve through practice.

Children of 9 and 10 years show improved motor development. As their strength and endurance increase, so does their interest in sports and other activities. By this age, children can actively participate in team sports (Fig. 7–3). They are now better able to understand rules and complex plays. Fine motor coordination improves, permitting them to learn to write as well as print.

Sexual Development

Sexual curiosity continues during this stage. Young school-age children ask many questions. Parents need to answer their questions honestly on the level of the child's

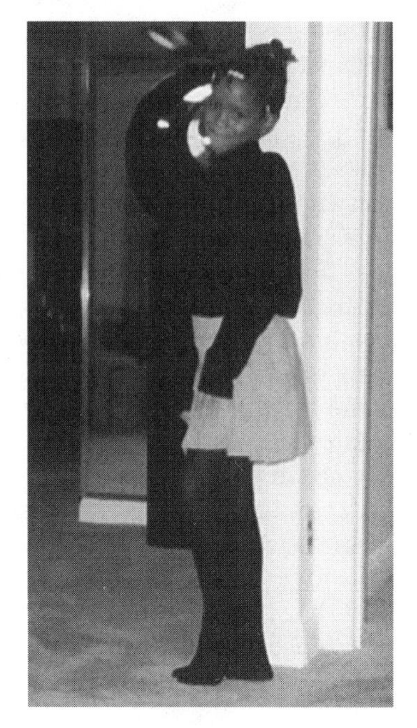

FIGURE 7–2
Eight-year-olds are more graceful and self-assured.

understanding, being mindful not to give more information than the child can digest or understand. Children learn much about their own sexuality and about the sexuality of others from their parents' behavior. Critical for the child's understanding are not only the details of sexual intercourse but how people feel about and treat others and how they handle issues of responsibility. School-age children need to learn about respect for other people's feelings and values even if they differ from their own. They should learn never to force another child to do something simply because it is what

FIGURE 7–3
Organized sports and peer interactions are important to the school-age child.

they themselves want to do. This teaches a valuable point without even focusing on sexual content.

Parents can use issues on television or lyrics of songs to form the basis of conversations relating to sexuality. Children should be asked questions like "Does that make sense?" or "How would you feel in that situation?" Parents need to provide all the information children need and must keep the lines of communication open so that their children will be able to make future decisions regarding sex.

Freud describes this period as **latency**, a time when sexual energies are relatively dormant. During this stage children are more involved with cognitive skills and learning than with sexual concerns. Because this is sometimes thought of as a period of "homosexuality," their peer relationships are mainly with children of the same sex. The ability to establish meaningful, caring relationships at this stage helps children prepare for caring relationships in adulthood.

Psychosocial Development

Industry

Erikson believed that school-age children are able to see themselves as producers. Thus he viewed the primary task for this stage of development as **industry**. Children at this time are more focused on the real world and see themselves as part of a larger group, allowing them to accomplish more and get along better with others. Their motivational drive increases and they gain satisfaction from their accomplishments. Schoolwork takes on a great deal of importance to children of this age. They often set very high standards of achievement in their academic endeavors. When they fall short of their goals, they may be very disappointed and develop a sense of failure. Some children become very upset if they do not make 100 percent or receive praise from their teachers for their efforts. For this reason it is very important to use positive reinforcement as motivation for learning.

Children of 6 to 7 years of age are full of energy and anxious to try new skills. Many begin new projects but don't have the patience or attention span to see things to completion. For example, a mother and child may begin to bake cookies, but halfway through the task the child loses interest, leaving the mother alone to complete the task. At the beginning of this stage children need immediate gratification for their work efforts. They are in a hurry to finish what they start and proudly show it off to others. They need the praise and reward of others to help strengthen their self-esteem and motivate them.

Nine-year-olds can initiate a task and are motivated to see the task to completion. They know what is expected of them and are more likely to conform to win the regard of adults. By age 11, most children are capable of working on more complex projects and can accept a delayed reward. Praise still helps strengthen their self-esteem. Without reinforcement and praise, children may develop a sense of inferiority.

During this period, children begin for the first time to move away from the family toward peer relationships (Fig. 7–4). These relationships are generally numerous and of short duration. Most of the time they gravitate toward peers of the same sex and openly express dislike for the opposite sex. Among 7- to 8-year-old children friendships become more intense and serious but are still mainly with children of the same gender. These friendships are made up of several children who have common needs and interests. They frequently form a strong friendship or best friend. Heroes or idols may

FIGURE 7–4
School-age children move toward peer relationships.

be worshipped and fantasized about by children of both sexes. Thinking and behavior become more complex. Activity levels vary greatly, with some periods of quiet sitting and other periods of high energy and activity.

School-age children are frequently engaged in rivalry with their siblings and often wind up in tears, still wanting things to go their way. They keep track of whose cookie has more chips in it or who sat in the front seat of the car during the last outing. Their words, anger, and level of competition seem out of proportion to the issues. For example, they often say, "I wish you were dead" to a brother or sister who has the favorite chair. Jealousy is a common emotional expression that may intensify when the child enters school, leaving younger siblings at home with undivided parental attention. Another form of this emotion occurs when the child sees peers as being more accomplished. Although at all ages children need love and affection, boys of this age tend to feel that they are too old to be kissed or hugged and may resent their parents' use of endearing terms or displays of public affection.

Family relationships appear to be less important to school-age children than their new peer relationships. Inside the home they often express negative feelings and are openly hostile toward family members, whereas outside the home they firmly defend, support, and even boast about their family members. Friendships are very important and are the cornerstone of the school-age child's social world. The patterns and interpersonal relationships that are learned at this stage continue into adulthood. School-age children are able to develop reciprocal relationships with their peers. These relationships are based on genuine feelings and appreciation of the others person's unique qualities. It is common for children to establish intimacy with their friends and to share their possessions, as well as their innermost secrets and feelings. Toward the end of this stage friendships become more intense and serious but are still mainly with children of the same gender.

Privacy becomes very important to children of this age group. They want their belongings and valuables to be off-limits to others. They also want privacy in self-care activities and appear modest and shy. For example, when shopping for new clothes with a parent, the child may insist on going into the dressing room alone. These feelings should be respected and not ridiculed.

It is not uncommon for children at this age to have an exaggerated fear of physical harm to themselves and to members of their family. Some of these fears may be increased by watching violence on television or in the movies. To help reduce these fears, children should be given realistic reassurance and their exposure to violent programming should be limited. At this age they may spend time worrying about issues such as divorce, illness, and dying. As children move toward adolescence, they tend to become more nonverbal about their worries, keeping things to themselves.

Preschool children recognize that money can buy things, but their concept of money is yet unclear. For example, they may believe that coins are worth more than paper money and that nickels are more valuable than dimes because they are larger. School-age children begin to place importance on money and possessions. Parents should establish with their child a predetermined amount of money as an allowance and use the allowance to help teach the child how to handle money. School-age children are also now capable of working at small jobs; in fact, many ask, "How much will you pay me?" before doing a simple household chore. Allowances and home chores should be kept separate. Children should be taught that they are expected to help around the home simply because they are part of the family unit. This teaches them responsibility within the family setting, and these principles can later be applied to the larger environment.

During this period there are often conflicts regarding the child's personal hygiene and other home-care activities. School-age children frequently have to be reminded to bathe and change their clothes. They may spend a great deal of time in the bathroom but may emerge no cleaner than when they went in. By the end of this stage children frequently leave their room a mess, suggesting that puberty and the teen years are rapidly approaching. See Box 7–1 for hints on preparation for puberty.

Emotions have a wide range of expression, depending on the child's chronological age and psychological maturity. In the early part of this stage of development children may use simple emotions to express their feelings. Tears, for example, are still used but are quickly seen as a babyish form of expression. Some children are shyer than others. Some have many fears or worry about their social acceptance or their school performance. See Table 7–1 for a list of common school-age fears. Children may express anger, a powerful emotion, in different ways (Fig. 7–5). Some may be negative or sulk, others may withdraw or refuse to speak, and still others may be openly disagreeable and hostile. Anger may represent the child's frustration and need for independence.

BOX 7–1

Preparations for Puberty

1. Offer information and answer questions regarding puberty.
2. Expect adult appearance to precede adult behavior.
3. Promote positive self-esteem.
4. Treat puberty as a positive experience.

TABLE 7–1
COMMON SCHOOL-AGE FEARS

Age	Fear
6–7 years	Strange loud noises, ghosts and witches, being alone at night, bodily injury, school
7–8 years	Dark places, catastrophes, not being liked, physical harm
8–9 years	Failure in school, being caught in a lie, divorce or separation of parents, being a crime victim
9–11 years	Becoming ill, heights, pain, evil persons

Discipline

Discipline continues to be an important need for children of this age group. It teaches children boundaries and helps to set limits to their behavior. Children also need a certain amount of freedom to explore. The proper amount of discipline is crucial. Too much may lead to acting-out behavior, with the child attempting to prove and assert himself or herself. On the other hand, too-lenient parental control may lead to insecurity and doubt. Children need adequate praise and rewards to help reinforce desirable behavior.

School-age children are usually able to take responsibility for their room and possessions. They can be given small jobs around the house as part of their chores. These jobs help give children a sense of importance within the family structure and help teach them responsibility.

Special Psychosocial Concerns

Television violence. Studies have indicated that violence on television can have an adverse effect on young viewers. Even children who have no problems dealing with aggression have been shown to become more aggressive after watching violent television. The belief that television portrays real-life events further complicates the school-age child's distorted views. It is not uncommon for children of this age to imitate and

FIGURE 7–5
Anger is expressed in different ways.

idolize cartoon or other characters from television shows. Parents can help children choose shows to watch and limit their viewing of excessively violent programs. In addition, parents need to discuss values and practice nonviolent behavior.

"Latchkey" children. Approximately one fourth of all elementary school children care for themselves for a short time after school. In some inner-city areas there may be after-school programs, but in more rural areas children are more likely to be at home unsupervised. It is difficult to generalize or predict how these "latchkey children" are doing. Some of them go straight home, report to their parents, and have a schedule of duties to complete. These children are less likely to get into trouble than those who are unsupervised and just "hang out."

Helpful Hints: Latchkey Children

- Arrange for the child to report to a specific place.
- Have the child check in with an adult at a specific time.
- Teach the child how to respond to emergencies.
- Designate a neighbor or close-by adult whom the child can call if he or she is worried or concerned.
- Rehearse how the child should answer the telephone and door.
- Arrange a schedule of activities for the child to follow.

Cognitive Development

According to Piaget, between ages 5 and 7, most children make the transition from the preoperational stage to concrete operational thought. School-age children have grasped the concept of conservation. For example, by 7 years of age the child understands that someone dressed in a costume is really just another human being, and not some alien being to be frightened of. Piaget believed that the school-age child develops causation, that is, understands the cause and effect of relationships. For example, 8-year-olds usually know that rain doesn't always cause thunder. According to Piaget, school-age children can also place objects in order according to size (seriation). Piaget also suggested that school-age children have improved conservation and classification skills. They learn to develop their conservation skills in an orderly sequence. That is, first they understand the concept of number, then the concepts of substance, length, area, weight, and volume. At this age they can recognize the relationship of a part to the whole and between sets and subsets. For example, 7- to 8-year-olds frequently have collections of stickers, baseball cards, books, videos, and so forth. These can be arranged and sorted according to several characteristics. Children of 8 to 9 years old can readily recognize and name different makes and models of cars. A car is not simply a Ford; it is a Taurus sedan. They can break down items into smaller parts and then reassemble them. Children at this age can enjoy working on more complex puzzles, assembling model cars, and so forth.

School-age children can take into consideration the views of other people, thereby widening their own perspective. Typically, they have an increased attention span and are less restless, which makes them better able to stay focused on activities for longer periods. Seven-year-olds are more serious than younger children. School-age children are very productive and adventurous. Their abilities have increased to allow a more organized style of thinking, which includes problem solving. They are better able to understand and follow rules.

School-age children can understand concepts such as time, space, and dimension with more clarity than they could at earlier stages. Their increased cognitive abilities enable them to master reading, mathematics, and science. During elementary school, children begin to learn how to tackle and solve a problem. School-age children have improved memory, which enhances their learning skills.

Children are often critical of others but may brag and boast about themselves. It is common to overhear them teasing, bullying, or using insulting comments when talking about others. Some children may assume the role of the victim and fall prey to a bully. Some children even find humor at someone else's expense. See Table 7–2 for parental guidelines for bullies and victims.

Starting formal education is a major accomplishment for this age group. School becomes the focus of the child's environment. In preschool the focus is on protection, play, and caring; in school the emphasis is on education and learning. School offers children the opportunity to establish themselves as individuals, separate from their parents and family. Children spend more time in school than anywhere else outside the home. This is where they try on different roles within a group setting, test their negotiation skills, and experiment with learning. Adjustment to school depends on the individual child, the home setting, and the school environment. For some, leaving home is very difficult and results in the onset of **school phobia**, which is an intense fear of going to school. The onset of various **somatic** (physical) complaints—stomachache, headache, or other unexplained pains—may be manifestations of school phobia and should be thoroughly investigated to rule out any underlying medical pathology or problems at school with the teacher, a bully, and so forth. Once the child has been medically evaluated, he or she should be treated if necessary, given support, and gently encouraged back to the school routine. School places several stresses on children. Just imagine finding a classroom by yourself, speaking in front of the class, or being reprimanded by a teacher. School, however, is also the place for rewards, recognition, and success.

In kindergarten, children learn to socialize and get along with their peers. Children at this age respond favorably to an authority figure, such as the teacher, and to praise and rewards.

TABLE 7–2
PARENTAL GUIDELINES FOR BULLIES AND VICTIMS

Bully	Victim
Teach to respect the rights of others.	Offer coping strategies.
Set clear, firm rules with regard to social behavior.	Encourage verbalization about incidents.
Teach and use negotiation techniques.	Encourage participation in activities that build self-esteem.
Set positive examples.	Praise child for achievements.
Praise desirable behavior.	Avoid intervening if at all possible.

During the early grades (1 through 4), children learn to follow routines and concentrate on specific tasks. Children of 8 and 9 years old are serious about their academic performance and are more ready to take responsibility for their learning and for their actions. They are concerned about being wrong and worry about being humiliated in front of their peers. They accept the teacher with unconditional respect.

Later, during the upper grades (5 through 8), learning becomes more of an independent task. Children of this age group are more judgmental and critical. Respect for teachers and authority wanes, giving the teacher less control in the classroom. Children need clear-cut rules, with an emphasis on fairness. Box 7–2 provides guidelines for a good classroom environment.

BOX 7–2

Desirable Classroom Environment

To create a desirable classroom environment, teachers should do the following:

Give ample praise.

Structure comfortable work areas.

Encourage sharing of responsibilities.

Stress academic achievement.

Use positive reinforcement.

Be a positive role model.

Promote open communication.

At an earlier age, homework was considered fun, but by 8 or 9 years of age children regard homework as something to avoid, rush through, forget, or put off. Homework and studying are the basis of many arguments between parents and older school-age children. Parents of young children are very involved with the mechanics of studying, assisting with assignments, and checking completed work efforts. Children who develop a routine for doing homework are generally more successful in school than those who do not.

By ages 9 to 11, most children would rather be at home or with their friends following their interests than in school. The children who thrive and succeed in school are usually more motivated and receive positive reinforcement from both parents and teachers. Poor performance in school at this time should not be ignored by parents or teachers.

Helpful Hints: Homework

- Try to make homework a pleasant experience.
- Set aside a specific time for homework.
- Make homework a top priority.
- Offer a lot of reassurance.
- Provide attention and contact with the child during homework sessions by offering a snack, a hug, or a reassuring touch.

Moral Development

Just as school-age children are making a transition in their cognitive abilities, they are also moving from one stage of moral development to the next. According to Kohlberg, most 6-year-olds are still likely to be at the preconventional level of moral thought. That is, they are primarily egocentric. They react to situations mainly to be rewarded or to avoid punishment or reprimand, without concern for other moral implications. Later, children move to the so-called conventional level of moral reasoning. They begin to make moral decisions based on what their family or others in society expect of them. They want to conform to what they believe will make them "good" girls or boys. By middle childhood, children can be counted on to act "the right way."

School-age children tell a different type of lie from younger children. They lie to improve their self-esteem and status and to win recognition. This lying is a type of bragging that helps them cope with new social pressures. These lies, if infrequent, should not be of concern to parents.

Helpful Hints: Lying

- Parents should set a good example. Do not ask your child to lie on the phone, for example, saying that you are not home when you are but don't feel like talking.
- Give the child permission to tell the truth by listening and not setting excessive punishments.
- Admit your own mistakes.

Toward the end of this stage, at about age 11, children begin to balance their self-interests and needs against what they know to be right. They also begin to consider what is fair to others. Kohlberg describes their concern for others as the beginning of **reciprocity**.

The 11-year-old child has a need to be trusted. Trust reinforces the child's concept of self-worth. At this point children demand loyalty from their friends and reciprocate this feeling. School-age children also learn self-regulation and control of their behavior.

Some studies have supported Kohlberg's views on moral development and have indicated that children's moral thinking is stimulated when they are involved in ethical discussions or in decision making. Other studies suggest that moral development is different for girls and boys: Girls' moral reasoning focuses on considering and preserving human relationships, whereas boys base moral decisions on protecting and defending others. Regardless of which theory one ascribes to, it is important to keep in mind that moral reasoning, like cognition, develops gradually, with some overlapping at each stage. A child's moral code is based on his or her parents' teachings and behavior. Once children internalize a moral code, they use it to judge the actions of others. Box 7–3 gives an example of a moral exercise.

Communication

Language improves so that school-age children are able to communicate more effectively with others. The ability to use language is important and enhances

BOX 7–3

Moral Exercise

Read the following exercise to the class. Ask the children how they would react in similar circumstances. Discuss the responses with the class. One day two brothers are walking home from school and are approached by a classmate, who gives the older boy a "joint" (marijuana cigarette) to smoke. The younger brother watches his brother accept and begin to smoke the cigarette. Despite his younger brother's protests, the older boy tells him not to tell anyone what he has just seen. He reminds the younger boy that he must keep this to himself or risk losing his (the older brother's) respect and friendship.

What should the younger brother do: Keep the information to himself or share it with his parents and risk upsetting his sibling?

socialization and group belonging. Earlier forms of communication, which may have been more limited to gestures and crying, are now socially unacceptable. The unacceptability of these forms helps to reinforce language acquisition.

School places an emphasis on building vocabulary, proper grammar, pronunciation, and sentence structure. School-age children can use nouns, verbs, and adjectives in their sentences. For the most part, they use the proper tense in speech. Their sentences describe their feelings, thoughts, and point of view. The use of swear words or slang increases as a result of peer influence. These words help to express their emotions and give them a sense of importance. A secret language is sometimes used by this age group to transmit messages back and forth to one another. Not only does this give children a degree of privacy; it also enhances their sense of group belonging. See Table 7–3 for a schedule of language development.

NUTRITION

For continued growth of the musculoskeletal system, nutritional requirements during this stage include an adequate intake of calories from proteins, carbohydrates, fats, vitamins, and minerals. Calcium is especially important at this time to allow the building of dense bones. Caloric requirements vary from child to child based on body size, activity, and metabolism. According to the food pyramid shown in Chapter 5,

TABLE 7–3
LANGUAGE DEVELOPMENT

6 years	Has vocabulary of 3000 words, understands meaning of complex sentences, can read
7 years	Tells time, prints well
8 years	Writes as well as prints
9 years	Describes objects in detail, writes well
10–11 years	Writes lengthy compositions, begins dictionary skill, develops good grammar

TABLE 7–4
SIGNS OF GOOD NUTRITION

Attitude	Alert, energetic
Height and weight	Normal range
Skin	Smooth, moist, good color
Hair	Shiny
Eyes	Clear, without circles
Teeth	White, bright, straight, and without discoloration
Gastrointestinal	Good appetite and proper elimination
Musculoskeletal	Well-developed, firm muscles and good posture
Neurological	Good attention span

children during this stage need two to three servings of meats, two to three cups of milk, four to five servings of vegetables, and four to five servings of breads and cereals per day. Table 7–4 lists indicators of good nutrition. The child's food preferences result from cultural influences, family preferences, and peer influences. By the time children enter school, their food habits are well established. These habits can be further influenced by their increased exposure to different cultures.

Important considerations in planning meals for the school-age child include maintaining weight within normal limits and avoiding a diet high in cholesterol. The recommended daily intake of cholesterol is 300 mg or less per day. Between 3 and 25 percent of school-age children have increased blood cholesterol levels, increasing their risk for cardiovascular disease. To keep cholesterol intake down, provide a diet that is low in **saturated fats**, that is, fats from animal sources, such as meat and dairy products. Both blood cholesterol levels and weight should be monitored at the child's routine health checkups.

Breakfast is one of the most important meals of the child's day. It should supply children with one fourth to one third of their daily nutritional needs. Children should not be allowed to skip breakfast before going to school. Lunch programs often supply children with the noontime meal. Parents should monitor the school lunch program to make absolutely certain that it meets the standards and nutritional needs of the growing child.

Food fads and new habits begin at this time. Excessive sweets and caffeine should be avoided as they may interfere with concentration and cause hyperactivity. Children should be encouraged to become involved in shopping for, preparing, and serving food and cleaning up after meals.

SLEEP AND REST

Sleep routines should be well established by this stage. The average 6-year-old needs about 12 hours of sleep, whereas the 11-year-old needs about 10 hours. The 6-year-old can tire easily and become very irritable. For this reason parents need to supervise activities and plan for quiet, restful activities to prevent overexhaustion. Some children may even benefit from a short afternoon nap. If children are chronically tired during school hours, their academic performance and social relationships will suffer.

Sleep for some younger children may continue to be restless and interrupted by nightmares. As the child grows older, nightmares usually decrease. They also become less terrifying and real as the child begins to be able to separate fantasy from reality.

PLAY

The focus of play illustrates the movement that school-age children make from pretending and fantasy toward reality and concrete thinking. School-age children are full of energy and willing to learn new skills. The style of play for this age group permits the use of cooperation and compromise. Many younger children are not yet ready for competitive activities and become upset at losing. Parents must be careful not to push their child into competitive play before the child is actually ready to handle this kind of demand.

Most 9- to 11-year-olds are involved in many play activities. They are usually very competitive and active. Children are now able to learn and follow rules and regulations. The style of play for this period of development is referred to as team play. **Team play** usually involves groups of the same sex and may be competitive in nature. Boys are likely to be better than girls at running, jumping, and throwing. This may stem from lack of opportunity and encouragement. Efforts are now being made to emphasize and fund girls' sport activities to the same degree as those for boys. This may help to enhance girls' sports perfomance and competitiveness. Much of the play of both sexes is based on group or team activities.

Children are chosen by their peers to participate on a team based on their ability to excel in the skills required to play the game. Being selected first helps validate the child's self-esteem and worth. By this stage children have learned that practice helps to improve their skills. Skill and motor coordination determine the child's overall performance and acceptance by others. Organized team sports such as Little League or team swimming may appeal to this age group. Competition now serves as a motivating force to practice and improve.

While most of the play during this stage involves sports and athletic activities, some children prefer sedentary activities such as crafts and board games. Collections of all sorts (cards, little boxes, cars, stickers, dolls, and other items, some of little or no value) are very popular at this stage. Ownership seems to have greater importance than actual monetary value.

SAFETY

The leading cause of accidental deaths continues to be motor vehicle accidents. State regulations for car safety should be carefully observed. Children should be taught to use seat belts whenever they are passengers in automobiles and school buses.

The increased motor skills of the school-age child help contribute to the increased numbers of accidents and injuries. An additional factor is that most children at this stage are ready to attempt any new skill with or without practice or training. For the most part they are much less fearful than they were during earlier years. To minimize danger from biking accidents, skating, or skateboarding, children should be encouraged to wear helmets and protective gear when engaged in these popular activities.

Many bone and muscular injuries occur during this stage, because of the child's physiological development and increased participation in sports. During school age the **epiphyseal cartilage**, or bone end, is the site for future bone growth. These points are weak and become potential sites for fractures. Final bone **ossification**, or hardening, occurs at puberty.

Helpful Hints: Safety

- Wear protective equipment.
- Observe traffic signals.
- Practice water safety: learn to swim, and never swim alone.
- Use the buddy system when walking to and from school.
- Never talk to or accept rides from strangers.
- Always follow your instincts and avoid peer pressure.

HEALTH PROMOTION

In general, the school-age period of development is considered healthier than earlier periods. It is common to see a slight rise in the incidence of upper respiratory tract infections when children first enter school, probably because of their exposure to many children. Once their immune system adjusts, they are able to resist many infections. As their organs continue to mature, there is less risk of ear infections, febrile seizures, and dehydration.

School-age children continue to need supervision in hygiene and daily care. There may be a higher incidence of urinary tract infections in girls related to their anatomy and daily toileting practices.

Before beginning school and every year thereafter, children should get their eyes checked. Most visual problems can be corrected with glasses and retraining exercises. Adequate lighting is important to help maintain proper vision. Regular hearing tests should be scheduled to determine baseline hearing levels. Parents should instruct children about avoiding exposure to excessive noise, which may damage the ear and lead to hearing loss.

School-age children need adequate exercise to help develop strength and muscle endurance. Poor posture can indicate fatigue or a minor skeletal defect. Children should be screened for **scoliosis**, the abnormal lateral curvature of the spine, during school and during their physical examinations. Have the child bend over and examine the lumbar thoracic region for unequal curvature. This condition is seen more frequently in girls than in boys. Early recognition can lead to prompt treatment and correction with exercises or braces.

Yearly checkups should also include a urine examination for infection and diabetes mellitus and blood tests for iron-deficiency anemia and cholesterol levels. Blood pressure screening should be instituted at this time. Other routine measurements include continued monitoring of weight, height, and growth. A part of health promotion must include nutritional guidance. By school age all of the primary immunizations have been completed and children need to receive boosters to help maintain immunity. They normally receive a tetanus-diphtheria booster every 10 to 14 years. Refer to Chapter 4 for the immunization schedule.

Special Health Concern: Substance Abuse

School-age children are easy prey to the effects of substance abuse through peer pressure. Some may gravitate toward alcohol, tobacco, or drugs based on a familiarity with such items in their home. Parents need to model desired behaviors and offer information and guidance to prevent substance abuse in their children. For a complete discussion of substance abuse, see Chapter 8.

SUMMARY

1. The period of development known as school age, the middle years, or late childhood is characterized by slow, consistent growth.
2. This stage begins with entrance into formal school and ends with the onset of puberty.
3. Five important accomplishments occur during this stage: (1) growth becomes slow and steady; (2) children move away from the family toward peer relationships; (3) children become less self-centered and more goal-directed; (4) deciduous teeth are replaced by permanent teeth; and (5) sexual tranquility replaces sexual curiosity and preoccupation.
4. The permanent teeth appear in the same order as the deciduous teeth.
5. Initially, school-age children experience an increased number of respiratory infections because of their increased exposure to other children. Once the immune system matures, this becomes a relatively healthy period of development.
6. Heart rate and respiratory rate slow, whereas blood pressure readings increase. Changes in the nervous system permit expanded cognitive processes.
7. There is marked improvement in existing gross and fine motor skills, permitting children more independence.
8. This period is sometimes called latency, a time when sexual energies are dormant. Because peer relationships at this stage are mainly with children of the same sex, this stage is thought of as a period of "homosexuality."
9. According to Erikson, the task for this stage of development is industry. Children are now capable of focusing on reality, and they gain satisfaction from their accomplishments.
10. Friendships are very important and are the cornerstone of the school-age child's social world. During this stage children are able to develop reciprocal relationships with their peers. It is common for them to establish intimacy with their friends and share their feelings and possessions.
11. Discipline teaches the child boundaries and helps to set limits to behavior. Too much discipline may lead to acting-out behavior; insufficient discipline may lead to insecurity and doubt.
12. Television violence may adversely affect many school-age children. For this reason parents should supervise children's choice of programming.

13. About one fourth of school-age children are at home alone while parents are at work. These "latchkey children" need special guidelines to follow while alone.
14. According to Piaget, in this stage children move from the preoperational level of cognitive development to concrete operational thought, which permits organized thinking and the ability to understand and follow rules.
15. School becomes a major focus of the child's environment. Unlike the experience in preschool, when the focus was on protection, play, and nurturing, the emphasis is now on education and learning.
16. According to Kohlberg, school-age children are at the preconventional level of maturity. Reciprocity is the concern for others that school-age children develop. Trust and loyalty are demanded of friends. The child's moral code is based on the teachings and actions of the parents.
17. Language improves, enabling children to communicate more effectively with others.
18. School places emphasis on the development of vocabulary, grammar, pronunciation, and sentence structure.
19. The nutritional requirements during this stage include an adequate diet of nutrients necessary for growth of the musculoskeletal system. Food preferences result from cultural, family, and peer influences. Breakfast is one of the most important meals of the child's day. It should not be skipped and should supply one fourth to one third of the daily nutritional needs.
20. The average 6-year-old needs about 12 hours of sleep, whereas 11-year-olds need only 10 hours of sleep each night. Inadequate sleep can produce irritability and interfere with the child's academic and social relationships.
21. School-age children play and carry out most self-care activities independently. They learn to write, draw, and dance and develop many other creative hobbies. In the home, children are able to take responsibility for their possessions and like to earn money for small jobs.
22. The style of play for this stage of development is referred to as team play. At this time that children are able to learn to follow rules and regulations. Most play occurs in same-sex groups and is competitive in nature.
23. The leading cause of accidental deaths continues to be motor vehicle accidents. Children should be instructed to use seat belts whenever they are passengers. Other causes of increased injuries are related to the school-age child's natural tendency to attempt new skills without help, supervision, or training.
24. Children of this age need adequate exercise to help develop muscle strength and endurance.
25. School-age children should be screened for abnormal curvature of the spine known as scoliosis. They need to receive boosters to maintain their immunity.
26. As school-age children's organs continue to mature, they are better able to resist infections and tend to recover more rapidly.

CRITICAL THINKING

Helen Lightbourne is the mother of 7-year-old Heather, her first-born. Ms. Lightbourne expresses concern to the pediatric nurse that Heather appears to have two loose teeth. Her specific concern is the management of the loose teeth.

1. What information should the nurse plan to share with Ms. Lightbourne about the expected pattern of tooth loss?
2. How would you instruct Ms. Lightbourne to care for Heather's primary teeth?
3. What common complication should she be alerted to as the permanent teeth erupt?

Multiple-Choice Questions

1. The change in posture typical of school-age children is due to:
   a. Tightening of the ligaments that support the long bones
   b. Lengthening of the musculoskeletal fibers
   c. A shift in the center of gravity
   d. Flattening and broadening of the rib cage

2. School-age children experience an increase in blood pressure because of:
   a. Decreased cardiac muscle strength
   b. Reduction in the capacity of the atrium
   c. A slight increase in hemoglobin
   d. Development of the ventricles

3. Stress experienced by school-age children may be manifested as:
   a. Ritualistic behavior
   b. Magical thinking
   c. School phobia
   d. Egocentric thinking

4. According to Erikson, the psychosocial task for school-age children is known as:
   a. Trust
   b. Industry
   c. Inferiority
   d. Initiative

5. During the school-age period, interest in sexual activity:
   a. Peaks
   b. Is dormant
   c. Directs the child's actions
   d. Involves heterosexual relations

6. The number of hours of sleep that the average 6-year-old needs is:
   a. 10
   b. 12
   c. 8
   d. 15

Suggested Readings

Ahmann, E, and Bond, NJ: Promoting normal development in school-age children and adolescents who are technology-dependent: A family-centered model. Pediatr Nurs 18(4):399–405, 1992.

Cillessen, A, van Ijzendoorn, HW, van Lieshout, C, and Hartup, W: Heterogeneity among peer-rejected boys: Subtypes and stabilities. Child Dev 63:893–905, 1992.

Brazelton, B: Kids and stress: How to help a child who worries too much. Family Circle, March 12, pp. 32–34, 1996.

Bussey, K, and Bandura, A: Self-regulatory mechanisms governing gender development. Child Dev 63:1236–1250, 1992.

Edelman, CL, and Mandle, CL: Health Promotion throughout the Lifespan. Mosby–Year Book, St. Louis, 1990.

Newman, J: Conflict and friendship in sibling relationships: A review. Child Study J 24(2):119–147, 1994.

Rogers, AY: Acceptability of time-out procedures for schoolage children: Evaluations by direct care staff and students in child development and child care. Child and Youth Care Forum 21(3):195–208, 1992.

Schuster, CS, and Ashburn, SS: The Process of Human Development: A Holistic Life-Span Approach. JB Lippincott, Philadelphia, 1992.

Socha, TJ, and Kelly, B: Children making "fun": Humorous communication, impression, management and moral development. Child Study J 24(3):237–251, 1994.

Wong, DL: Essentials of Pediatric Nursing. Mosby, St. Louis, 1993.

Puberty and Adolescence

Chapter 8

Chapter Outline

Physical Characteristics
- Puberty
  - *Height and weight*
  - *Development of sex characteristics*
- Adolescence
  - *Height and weight*
  - *Muscle and bone development*
  - *Development of other body systems*

Vital Signs

Developmental Milestones
- Motor Development
- Sexual Development
  - *Sexually transmitted diseases*
  - *Teen pregnancy*
  - *Rape*
- Psychosocial Development
  - *Puberty*
  - *Adolescence*
  - *Discipline*
- Cognitive Development
- Moral Development
- Communication

Nutrition

Sleep and Rest

Exercise and Leisure

Safety

Health Promotion
- Depression
- Substance Abuse

Summary

Critical Thinking

Multiple-Choice Questions

Suggested Readings

Puberty and Adolescence

Key Words

adenohypophysis
adolescence
ambivalence
anorexia nervosa
apocrine glands
bulimia
depression
ejaculation
emotions
estrogen
gonads
larynx
menarche
ova
ovaries
penis
preadolescence
primary sex characteristics
progesterone
puberty
secondary sex characteristics
sexually transmitted diseases (STDs)
scrotum
sperm
testes
testosterone

Learning Objectives

At the end of this chapter, you should be able to:

- List four physical changes occurring in puberty.
- List four physical characteristics of adolescence.
- Describe three developmental milestones seen during the adolescent period.
- Describe the primary psychosocial task of adolescence as identified by Erikson.
- Describe the cognitive level of functioning during the teenage period of development.
- State how teens develop moral reasoning.
- List three factors that help to promote wellness in the teen.
- Describe three special concerns that may adversely affect adolescent health.

The period known as **puberty** or **preadolescence** is a time of rapid growth normally commencing between ages 11 and 14 and taking an average of 2 years to complete. It is marked by the development of secondary sexual characteristics. Puberty ends (and adolescence begins) with the onset of menses, or **menarche**, in girls and the production of sperm in boys. The growth patterns affecting the onset of puberty are influenced by several factors, including heredity, climate, nutrition, gender, and socioeconomic status (Fig. 8–1).

Four major changes associated with the pubescent period are:

1. Rapid physical growth
2. Changes in body proportions
3. Development of primary sex characteristics (sex organs)
4. Development of secondary sex characteristics

The term **adolescence** is from Latin and means "to grow and mature." It refers to a transitional period, which begins with sexual maturity and ends with cessation of growth and the movement toward emotional maturity. This period of development bridges the gap between dependence and independence, or childhood and adulthood. Adolescents need to accomplish the tasks that help prepare them for adulthood. The major characteristics of adolescence include:

- Stormy emotions
- Feelings of insecurity
- Introspection
- Experimentation and learning
- Testing values and beliefs

FIGURE 8–1
The age of puberty varies for different children.

PHYSICAL CHARACTERISTICS

Puberty

Height and Weight

Puberty is the second greatest period of rapid growth, after the prenatal period. This growth spurt occurs in girls earlier than in boys. Height increases 20 to 25 percent. Boys grow 4 to 12 in (10 to 30 cm) and girls grow 2 to 8 in (5 to 20 cm) during this period. Increases in weight follow increases in height. These weight changes are related to an increase in fat, bone, and muscle tissue. Boys gain between 15 and 65 lb (7 and 30 kg) and girls gain 15 to 55 lb (7 to 25 kg). Different parts of the body grow at different rates, making the whole seem temporarily out of proportion. The bones grow longer and change in shape. The trunk begins to broaden at the hips and shoulders.

Development of Sex Characteristics

The **primary sex characteristics** are the **gonads**, or sex glands. The gonads are present at the time of birth but remain functionally inactive until the onset of puberty. The maturation of these glands is influenced by the **adenohypophysis**, or anterior lobe of the pituitary gland. The pituitary gland secretes a hormone that stimulates the gonads. In boys, the male gonads, or **testes**, are located in a sac called the **scrotum**, found outside of the body. The testes produce male sex cells, or **sperm**, and the male sex hormone **testosterone**. **Ejaculation**, the release of sperm, indicates that the testes are functionally mature. In addition, the **penis**, or male sex organ, grows in length and circumference.

The female gonads or sex glands are the **ovaries**, located in the pelvic cavity. Their primary function is to produce the female sex cells (**ova**, or eggs) needed for reproduction and the female sex hormones **estrogen** and **progesterone**. The onset of the menstrual flow, which is called **menarche**, indicates that a girl is capable of reproduction. The menstrual flow is a monthly discharge of blood, mucus, and tissue from the uterus. It lasts from puberty until menopause. The usual monthly cycle is every 21 to 24 days, with each monthly period or discharge lasting an average of 5 days. Average blood loss with each period is 30 to 60 mL. Some girls and women experience headaches, cramps, swelling, and irritability before and at the onset of their period.

The whole menstrual process is an emotionally charged event. Attitudes toward menstruation are assimilated from cultural and personal experiences. Education regarding menarche should begin during the school-age period in the home setting. Adequate preparation leads to a more positive initial experience. The primary concerns of young girls are related to hygiene, preventing clothes from getting soiled, and embarrassment. In addition, girls at this stage need to be informed about restrictions, activities, and taboos. Several misconceptions about menstruation may still exist and need to be openly explored with young girls.

Secondary sex characteristics play no direct role in reproduction but appear at this time. Initially, pubic hair is sparse and lightly pigmented; it then becomes darker, coarse, and curly. In boys axillary and facial hair appear after the pubic hair growth. Boys' skin thickens, and hair appears on the arms, legs, shoulders, and chest.

In both boys and girls, the sebaceous glands produce oil and become larger and more active. This increased activity may be related to the appearance of acne (pimples) seen in many adolescents at this time. The **apocrine glands** (sweat glands) in the armpits and groin become larger, producing a characteristic odorous secretion.

In boys, the **larynx**, or voice box, and the vocal cords increase in size, resulting in a deepening of the voice. Changes in the distribution of fat and increase in the width and roundness of the hip and pelvic bones are secondary sex characteristics occurring in girls. Breast development in girls follows an orderly sequence, resulting in the increase of fatty tissue and the maturation of the mammary glands. Hair also appears in the groin and the axillae. See Table 8–1 for the signs of puberty.

Adolescence

Height and Weight

The rate of physical growth slows down after puberty. In girls growth in height ceases between 16 and 17 years. Boys continue to grow in height up to 18 to 20 years. During adolescence body proportions are similar to those of the adult.

Muscle and Bone Development

Muscle strength and endurance increase along with a growth in the muscle size. Some adolescents complain of muscle soreness and fatigue with increased activity. Adolescents may at first be awkward as a result of the patterns of muscle growth. But by the end of this stage they should have good muscle development and coordination. Motor capabilities improve with practice and training. Posture may be poor, evidenced by slouching. This may be further complicated by a common condition causing lateral curvature of the spine, known as scoliosis. As discussed in Chapter 7, this is more commonly seen in girls than in boys.

Development of Other Body Systems

The weight and volume of the lungs increase, causing a slowing down in the respiratory rate and an improvement in lung performance. Exercise helps to improve both cardiac and respiratory function.

TABLE 8–1
NORMAL PUBERTY AND ADOLESCENT DEVELOPMENT

Girls	Boys
10–11 years: rapid growth spurt, breast development, appearance of pubic hair	11–12 years: growth of the testes, scrotum, and penis; appearance of pubic hair
11–14 years: first menstrual period	12–13 years: rapid growth spurt
12–13 years: appearance of underarm hair	13–15 years: growth of underarm, body, and facial hair
	13–14 years: ejaculation
	14–15 years: deepening of voice

The stomach and intestines increase in size and capacity. Adolescents have an increased appetite and therefore require an increase in their daily food intake. Adequate food intake helps to meet the demands of their bodies. A common observation of adolescents is that they are always hungry and can consume enormous amounts of food at one time. The typical teenager will devour groceries as soon as they are taken out of the shopping bags.

At about 13 years of age, teens gain their second molars, and between 14 and 25 years their third molars, or wisdom teeth, appear. The jaw reaches adult size toward the end of adolescence.

VITAL SIGNS

Normal pulse range for this stage is between 60 and 90 beats per minute. The respiratory rate during adolescence is near that of the adult. Normal respiratory rate for adolescents should be in the range of 16 to 24 breaths per minute. Exercise at this stage produces an improved physiological response. The changes in the circulatory system include an increase in the size of the heart and in the thickness of the walls of the blood vessels. These changes result in an improvement in the pumping ability of the heart. There is also an increase in blood volume. In boys, greater force is needed to help distribute blood to the larger male body mass. This, in turn, causes an increase in blood pressure.

DEVELOPMENTAL MILESTONES

Motor Development

In the beginning of the adolescent stage, teens exhibit some clumsiness as a result of their rapid physical growth. The adolescent's motor functions are comparable to that of the adult. Eye-hand coordination now is markedly improved, allowing for good manual dexterity.

Sexual Development

Teens at first gravitate toward individuals of the same sex and ridicule those of the opposite sex. As their bodies undergo the physical changes of puberty, they suffer from heightened emotions, increased worries, and a lack of self-confidence (see Table 8–2). They are sensitive about the size of their body parts and readily compare themselves to their peers. Girls are preoccupied with the size of their breasts and boys are concerned about the size of their penis. These concerns continue into adolescence. From the time of sexual maturity, the teen can be sexually aroused to orgasm through self-stimulation. Masturbation is a normal part of sexual expression and has no harmful effects. But masturbation can result in feelings of anxiety and guilt if the adolescent is led to believe that it is shameful or unhealthy. Parents should respect adolescents' need for privacy and knock before entering their rooms.

The extent and age of onset of sexual behavior varies from individual to individual. Recent surveys indicate that sexual activity begins early in this country; some children are sexually active at age 10 or 11. Typical sexual behavior in the early dating period includes kissing, necking, and petting. Many teens engage in

TABLE 8–2
TEENAGERS' CONCERNS ABOUT THEIR CHANGING BODIES

Boys	Girls
Penis size: most boys compare penis size with peers; average length is 5–7 in (12–17 cm) when erect	*Breasts:* may not be exactly symmetrical; size varies; choose a well-fitting bra
Embarrassing erections: may occur at any time; try to think of something else to help it subside	*Menstruation:* usually begins between 11 and 14 years of age, lasts about 5 days, and occurs every 28 days
Morning erections: commonly occur during dreaming	*Menstrual discomfort:* to relieve, use heat, Advil or Tylenol, and moderate activity
Wet dreams: ejaculation may occur during dreaming	*Pregnancy:* becomes possible with onset of menstruation
Voice changes: the voice box enlarges and the voice gradually deepends	*Hygiene:* regular bathing or showering necessary especially during menstruation
Perspiration: apocrine sweat glands are highly active; wash daily and use a deodorant	Same concerns
Acne: skin blemishes more common—need daily cleansing. Topical skin-colored creams may help cover small blemishes	Same concerns

sexual intercourse. Studies indicate that girls sometimes become sexually active because of pressure or coercion, and that the first sexual encounter may be a great disappointment. Boys may also become sexually active because it is expected of them.

Sex is given high priority at this time in adolescents' lives. Girls traditionally set the limits on sexual interactions. A great deal of pressure to conform to the group's behavioral standards exists. Sexual uncertainties and confusion often arise; many teens give in to peer pressure or lie to their friends about their sexual adventures. By conforming, the individual may find group acceptance. Successful resolution to the search for identity is necessary during the teen years for the young person to find emotional sharing and intimacy.

Research indicates that teenagers acquire most of their information about sex from their peers rather from more authoritative sources. This may result in the transmission of incorrect information. Youngsters should receive sex education before they become teenagers. Young people need to be well informed about reproduction, their bodies, and the responsibilities of sexual behavior. Good sex education, which should include explicit information about the prevention of sexually transmitted diseases and unwanted pregnancy, enables adolescents to make responsible choices about their own sexuality.

Sexually Transmitted Diseases

The incidence of **sexually transmitted diseases** (STDs) is increasing for this population. The best deterrent is information: every teenager should be taught safe sex practices, whether they choose to be sexually active or not. STDs include chlamydia, trichomoniasis, herpes genitalis, gonorrhea, syphilis, and AIDS. Each of these diseases has its own cause, signs and symptoms, and plan of treatment, but all are spread through vaginal, oral, and rectal intercourse. Refer to Table 8–3 for a summary of sexually transmitted diseases.

TABLE 8–3
SEXUALLY TRANSMITTED DISEASES

Name	Symptoms
Chlamydia	May be asymptomatic or may have a yellowish vaginal discharge, painful or difficult urination, spotting between menses or after intercourse; may spread to other pelvic organs
Trichomoniasis	Thin, frothy, yellow-green vaginal discharge, vaginal itching, tenderness, redness, painful urination and intercourse
Herpes genitalis	Genital blisters, pain, swollen glands, vaginal discharge and itching
Gonorrhea	May be asymptomatic or may have a purulent yellow-green vaginal discharge; painful urination and intercourse
Syphilis	Painless ulcer (chancre) found on genitals, lips, or anus in early stage
Acquired immunodeficiency syndrome (AIDS)	Lethargy, weight loss, skin lesions, and fungal infections

Teen Pregnancy

Teen pregnancy is of social concern for a number of reasons. Although teenagers may be physically mature, they may not yet be emotionally mature enough to handle parenthood. Pregnancy and parenthood during the teen years interrupt plans, education, and usual activities. Teen pregnancy carries increased risks for complications for both mother and baby. Early recognition and medical supervision may help promote a more positive outcome. Young teenagers who become pregnant may need additional counseling and time to choose between their various options: abortion, adoption, or parenting. Their decision may have a serious impact on them for the rest of their lives.

Rape

Adolescent rape appears to be on the increase, and a large number of offenders are themselves adolescents. The exact numbers of rape cases are not clear because young women are the least likely to report these crimes. Not only are teens at risk for stranger rape, but they are at risk for date rape as well. Education for both male and female teens helps decrease myths about rape and provide preventive strategies.

Psychosocial Development

Psychosocial development is rapid throughout puberty and adolescence. Table 8–4 gives a review of teenage behavioral issues and how parents can deal with them.

Puberty

Several general behavioral characteristics are commonly seen during the puberty stage of development. Individuals enter this stage happy and slowly become negative

TABLE 8–4
TEENAGE BEHAVIORAL ISSUES

Rebelliousness, argumentativeness, or rudeness	Overlook what you can. Avoid confrontation. Be as tolerant as possible. Avoid seeing behavior as a rejection of parental love.
Need for privacy	Make certain that teen has own space. Understand teen's self-conscious behavior. Accept individual's need for some secrecy. Offer help, but step back if rejected. Keep open communications.
Dishonesty	Avoid overreaction. Reinforce reality. Have consistent principles.
Responsibility	Listen attentively. Expect maturity to be uneven. Encourage decision making and acceptance of responsibility.
Curfews	Set reasonable times. Allow for unexpected delays. Encourage frequent phone contact. Set good examples for teen.
Friends	Allow social life to center around the home. Accept friends without criticism. Try to get to know friends. Avoid showing open disapproval.

in their attitudes and interactions. The basis for some of this negativity is their growing self-consciousness. This self-doubt and worry are related to their changing bodies. Much of the behavior seen is influenced by an overall negative outlook. Youths at this age tend to spend more time by themselves and in their room than they did at earlier ages. Many move away from their earlier friendships and need to find their place in new group settings. Until this happens, they may be isolated and alone.

Social antagonism is demonstrated best by their interactions with family, peers, and society. In the family setting preadolescents are argumentative with their parents and jealous of their siblings. Their desire for independence becomes the root of conflicts with authority figures. They resent supervision and directions and see this help as a sign of their weakness and helplessness. This antagonism may extend into heterosexual interactions. Family relationships change dramatically during this stage of development. These changes produce turmoil and conflict. In the struggle for independence, the teen wishes to be free of restrictions and parental control. Chores, curfews, dating, telephone, money, driving, schoolwork, and friendships are some of the issues that spark disagreements between parent and child. Parents want teens to listen and conform to their regulations. Teens complain about feeling that they are not trusted. A common cry is, "Why can't I go or do like everyone else can?" Teens are often argumentative and critical of their parents' ways. Some teens act withdrawn and confide less in their parents, leaving the parents to feel cut out and removed. Teens often act as though they are embarrassed to be seen with their parents.

Helpful Hints: Living with a Teenager

- Allow for teens' privacy: provide them with a room of their own where possible.
- Recognize that teens are self-conscious and sensitive about their changing bodies.
- Don't expect to know everything about their thoughts and feelings because they may choose not to discuss everything with you.
- Keep the lines of communication open despite their demand that you leave them "alone."
- Try not to be too critical. Listen carefully.
- Recognize and praise their accomplishments.
- Show interest in their activities.
- Encourage them to bring their friends home.
- Set reasonable limits.
- Avoid arguments; exercise compromise.

Adolescence

Erikson described the primary psychosocial task for adolescents as the search for identity. The individual must answer the question, "Who am I?" Identity begins with a separation of the individual from the family. As they begin to separate, they start to explore and then incorporate ideals and values that will become part of their own self-concept. Before they can accomplish this fully, they test out and question these values and beliefs, comparing them to the beliefs of outsiders. Confusion, depression, and discouragement often accompany this period. Marked fluctuations appear in adolescents' moods, ranging from low self-esteem to feelings of grandiosity. They are more likely to experience increased physical symptoms at this time.

This is a difficult time for both adolescents and their families because they tend to blame their parents for most of their problems. The movement away from the family expresses the teen's need for freedom and independence. Complicating this need is their continuing need for parental love, support, and guidance. These conflicting needs for independence and dependence create what is known as ambivalent feelings. **Ambivalence** describes two opposing feelings about the same person or object. Adolescents are truly ambivalent about many issues: loving and hating their family, wanting freedom and needing supervision, wanting to be part of a group of peers and wishing to be left alone. Along with ambivalence, teens experience many different, sometimes conflicting emotions. **Emotions** are the expressed feeling tones that influence the person's behavior. The charged emotions characteristic of this period are caused by both the physical and hormonal changes that are occurring. The increased social pressures placed on this age group further heighten adolescents' emotional responses. Boredom is common during this development period. Individuals give up earlier forms of play activities, fearing that they represent babylike behavior. Daydreaming and fantasizing may occupy a great deal of their time alone. A sense of humor is generally present but often used at the expense of others. Name-calling and teasing peers and others seems to give them a sense of satisfaction. Although they are likely to tease others, they are usually not able to handle teasing that is directed toward themselves.

Emotions cover a wide range of expression. Some commonly expressed emotions are anger, fear, worry, jealousy, envy, and happiness. *Anger* can be very disruptive and destructive to relationships. It is often expressed when teens are denied privileges. They angrily complain that they "are being treated like children." Anger can also result when they are teased, criticized, or lectured. The manner in which they express their anger varies from individual to individual. They may sulk, withdraw, or have an angry outburst. *Fears* may be imaginary or real and are usually related to social situations or inner feelings of inadequacy. *Worries* stem from different issues related to school performance, vocational choices, relationships, appearance, and group acceptance. *Jealousy* may arise in their relationships, whereas *envy* is mainly related to social status and material possessions. *Happiness* occurs when the individual succeeds and feels at ease in different situations.

Many teens find some part-time employment. Work has many benefits for this age group. Work helps the individual develop knowledge and skills that can be applied in adult life. Work gives the adolescent a sense of belonging in the adult world. Furthermore, work teaches responsibility and provides a source of income. This teaches money management and principles of saving.

Ages 13 to 14 Years Young teenagers may hide their feelings from others and sit and sulk instead of opening up and discussing their feelings. They become openly more negative and hostile. Issues are seen mostly from their own point of view. This narrow perspective creates an attitude of intolerance toward others. Compromise is something that 14-year-olds find hard to do. Friends are very important at this time, with teenagers identifying more with their friends than their family. Boys tend to have small groups of friends, whereas girls usually have one or two best friends. Friendships create a very needed sense of belonging. Experimentation with clothing and hairstyles begins at this time. Door-slamming and harsh verbal outbursts are typical ways in which the 14-year-old deals with stress. Teens' sense of humor is based on their negative outlook. They readily insult or tease parents and siblings and may show disrespect to their teachers.

Ages 15 to 16 Years By this age teenagers are less self-absorbed and better at compromising. They are now more tolerant of other views. They can now think more independently and make more of their own decisions. Their curiosity and interests increase, allowing them to further develop specific skills relating to math, science, music, or sports. Thinking is abstract and can be extended into areas of discussion and debate. It is still likely to see them continuing to experiment with clothing, hairstyles, and attitudes. This experimentation helps them to shape their self-image.

Teens at this age often test their boundaries, pushing them to the limit. Risks are taken as they see themselves as impervious to danger. In fact, some teenagers believe that they are immortal. Socially, many teens are less shy and more adventurous. Many show an interest in travel with their clubs or organizations.

Dating begins around 15 to 16 years of age. "Crushes" are typically found in the early dating period. These are usually accompanied by strong feelings of attachment and what the individual believes to be love. These crushes usually last 1 to 6 months.

Physical attraction to an individual of the opposite sex is the immediate factor that draws the other party's attention. Dating is the major source of fun and recreation for the teenager. It helps establish a means of social status and recognition within the peer group. It also provides the individual with personal and social growth. For the teen who begins dating later than others or is less popular, the pressures of dating may lead to feelings of inadequacy and rejection. Depression or a profound feeling of unworthiness may result and must be closely monitored. At first, dating may be characterized by short-lived sexual relationships. Some teens are secretive about their dating and feelings, whereas others share their innermost thoughts with their closest friends.

Ages 17 to 19 Years At this point in development a sense of seriousness becomes more evident. Teens are now very involved with their own activities in school, at work, or with friends. Regardless of which activity they have chosen, the one common denominator is that the activity is based outside the home, keeping them away from the family for longer periods of time. At 17 to 19 years, teens are very idealistic. They like to work for a cause and follow the ideals they hold important and right.

Stress increases for this age, related to the many uncertainties about the future. As the stress level increases, so do temper outbursts. Difficulties continue between the individual and parents. Usually teens believe that they know more and are more in touch with the real world than are their parents. This leads to frequent discord and disharmony. By this age many teens have established more stable sexual relationships. They may have one serious boy or girl friend with whom they spend a good deal of time. In these later relationships the individual focuses on deeper traits such as: honesty, reliability, and a sense of humor. Sexual behavior varies with individuals based on earlier teachings and peer pressures. See Table 8–2 for teenagers' concerns about their changing bodies.

Peer relationships are very important at this stage. Peers share the same age, feelings, experiences, goals, and doubts in ways that parents can't. Friendships tend to develop along similar social classes and interests. Peers offer social and emotional fulfillment. Teens wish to have their peers recognized and accepted by their parents. The development of the self-concept is further influenced by the person comparing his or her own perceived appearance to that of peers. Real or imagined differences threaten self-esteem. A slight blemish or defect will be magnified and further weaken their confidence. See Box 8–1 for ways in which parents can promote self-esteem in their adolescent children.

Socialization is further developed through peer relationships. The social behavior of the adolescent changes from earlier patterns to resemble that of the social group to which they belong. The influence of the group on individuals depends upon the amount of intimacy and contact that they share. Teens may form cliques, crowds, and gangs. An important feature of these groups is that the individual must conform to the patterns or rules they determine to be socially acceptable. One of the strongest needs for a teen is to feel accepted by the group members. Perceived acceptance or the lack of it will influence the teen's behavior and attitude. Different personality traits emerge at this stage. The popular type feels secure, happy, and confident. The unpopular teen feels alienated, resentful, and antagonistic. The role of a leader falls onto the person having the most admired qualities.

BOX 8–1

Promoting Positive Self-Esteem in Adolescents

- Be positive.
- Recognize achievements.
- Be genuinely interested.
- Be sensitive.
- Encourage self-expression.
- Value opinions.
- Avoid belittling.
- Show respect.
- Encourage decision making.

Many demands are placed on adolescents at this stage. Society expects them to select a vocation and think seriously about their future. Impending graduation from high school causes teens to wonder if they should continue with their schooling or begin a job (Fig. 8–2). This question is of great magnitude for individuals who are unable to simply decide what to wear in the morning or whether to go out on a date or with a group of friends. Those who are unable to select a career may develop fear and self-doubt.

Other demands placed on this age group include the development of a value system and demonstration of socially responsible behavior. At the end of this stage the individual should have moved toward becoming more economically independent.

Toward the end of this stage many adolescents are able to bridge the generation gap by establishing close relationships with their grandparents (Fig. 8–3).

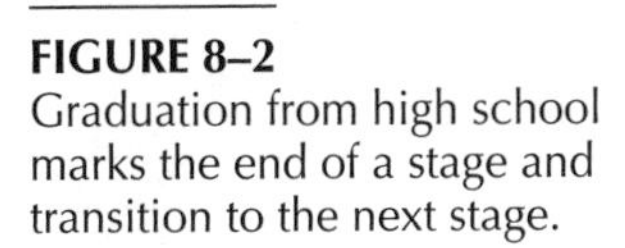

FIGURE 8–2
Graduation from high school marks the end of a stage and transition to the next stage.

FIGURE 8–3
Adolescents and older adults bridge the gap between generations.

Discipline

Discipline during adolescence is very important. Many of the conflicts between parents and teens are based on choice of friends and issues surrounding dating. Different parenting styles have different effects on the developing teenager (discussed in Chapter 2). The democratic style of parenting encourages youths to make decisions. Parents always have the right to approve or disapprove of expressed beliefs. This style of parenting best supports the child's developing sense of self. In the autocratic style of parenting the youth is not permitted free expression of feelings or views. Parents make decisions based on their own feelings and judgments. The teen is then expected to follow along with what the parents decide for them with little input. This style of discipline may hinder or slow the teen's growth process and moral development. In the laissez-faire parenting style the adolescent is left to decide what he or she wants or believes is best. Although most youths say that they want most to be free to decide, in this set of circumstances they feel ignored and unloved. It appears that feelings of independence and strength come easiest to those whose parents listen, explain, and make clear what is expected of the youth.

The use of "grounding" is appropriate for serious offenses by adolescents. This technique provides them with an opportunity to learn the consequences of their behavior. When grounded, the individual needs to be given specific jobs or household chores to be completed within a certain time frame. The parent must evaluate and praise the completed tasks. Grounding is effective when the youth generally follows the rules and shows an understanding of the consequences of the offenses.

Cognitive Development

During the adolescent period maturation of the central nervous system may lead to a shift from concrete thinking style to formal operational thought processes. Training and studies help the person progress from concrete thinking to more formal ways of reasoning. Some individuals move out of the concrete type of thinking while others without the proper training or motivation may not move beyond this point of cognitive development.

Formal operational thinking is conducted in a more logical manner. Some scientific reasoning and problem solving can be mastered at this point. Individuals are capable of looking at all possibilities. They can think abstractly beyond the present and imagine the sequences that might occur and the consequences of their actions. This form of thinking doesn't guarantee that they will make the right choices. Other factors such as peer pressure or the need to be accepted by the group or to look "cool" often have a greater impact than reason and judgment. Adolescents are able to analyze a problem, set up a hypothesis, collect evidence, and come up with possible solutions. In addition, they are more conversant on many more topics. Topics that hold their interest include politics, religion, justice, and other social issues.

School is at the center of teens' development. Most of their time is spent at school or in activities related to their schooling. Social skills, friendships, and peer interactions are of utmost importance to the teenager. Currently the law in most states requires teens to remain in school until the age of 16. Transition from middle school into high school may be both exciting and stressful for the teenager. Several factors complicate the stress of high school. Teens must come to terms with their changing appearance and developing self-image. Self-doubt may serve to complicate their ability to enter new relationships. New friends are sought out, with the possibility of joining many new groups and clubs. In high school teens are expected to be more independent and responsible for the learning process. This places more challenges and demands on them. Instead of having only one teacher for all subjects, they now have a teacher for each subject.

Many differences exist in the performance and academic achievement of today's teenagers. Several factors may determine the individual's success in high school, including socioeconomic background, family relationships, and peer and social pressures. Life after high school may include college near home or away, work, marriage and parenting, or some combination of these (Fig. 8–4).

FIGURE 8–4
For some adolescents, college begins the career path.

Moral Development

Cognitive development is a prerequisite for moral reasoning. Moral judgment is based on the earlier learned principles of right and wrong. Parents directly and indirectly influence the moral judgments of their children. Positive listening and empathy enable families to foster moral development. Teenagers must learn to make decisions for themselves and guide their own behavior according to learned standards. Society has determined that adolescents can no longer expect adults to guide all of their decisions.

During early adolescence teens are usually at the conventional level of moral development. They can follow rules and show concern for others. They express a strong wish to be trusted by their parents. Following this they progress to the transitional phase of moral development; they begin to question everything and everyone. This questioning places them in direct conflict with any person of authority, but it also helps them to gain autonomy from adults and begin to substitute and try out their own code of ethics. Teens look at rules and see many injustices, and feel that they have the right to change these rules. They believe that they can make a difference and want to get involved in social issues. They are always ready to take a stand on what they believe. Slowly they gain responsibility and show an understanding of duty and obligation.

Adolescents are further developing their spiritual awareness. They begin to question and compare religions. They can philosophize and think logically about religious doctrines. They speculate, search, and think about conflicting ideologies. During this period of awakening they may reject formal or traditional religious practices in favor of their own style of practice. Some move completely away from their family practice and may gravitate toward another less traditional ideology. A small number of teens may gravitate toward certain groups or cults.

Communication

Language skills and vocabulary increase during adolescence. Verbal communication is the means that adolescents use to make their thoughts and beliefs known. They will verbally argue or defend their ideas. Adults need to encourage their free expression and should listen and exchange opinions. This give-and-take type of relationship helps to foster teens' growth and sense of self.

In peer settings adolescents frequently develop a common language typical to their group, time, and culture. Having their own slang creates a sense of belonging for the teen and sets them apart from others.

NUTRITION

Because of the rapid growth that occurs during the adolescent period teens need an increase in calories, protein, minerals, and vitamins. The average caloric need for teenage girls is 2600 calories per day; for teenage boys it is 3600 calories per day. These needs are easily filled because the teen's appetite increases, allowing increased intake of food. Boys never seem to get enough food to keep them from feeling hungry. This age group also seems to enjoy food more than they did as children. The teenager's protein intake should be between 12 and 16 percent of the daily dietary intake. The increased need for calcium is necessary for skeletal and muscle growth and for the increased amount of total blood volume. Iron intake must be monitored after menarche because girls at that time may be more susceptible to iron-deficiency anemia.

Eating habits are affected by many factors. Time, pressures, and peers influence teen eating patterns. Fruits and vegetables are often passed over for other favorites such as meats and potatoes. Snacks are generally chosen for their accessibility and taste with little regard for nutritional value. Some youths favor milk and others move away from it toward carbonated drinks.

Dieting has become a national pastime; hundreds of fads and quick-weight-loss gimmicks are advertised. Current fashion trends may have added to the increased number of eating disorders. **Anorexia nervosa**, which affects a large number of adolescent girls, is the willful starvation that can result in weight loss of as much as 25 percent or more of a teen's body weight. This is more than an eating disorder. It is a complex emotional disorder that requires immediate medical attention. Anorexia accounts for a large number of deaths among adolescent girls. **Bulimia**, another eating disorder, is characterized by a series of binges followed by periods of purging or self-induced vomiting. This condition also warrants intense medical treatment. Obesity may also be listed as an eating disorder. Some teens are unable to maintain their desired weight and are in need of nutritional supervision to promote healthy weight loss.

There may be a slight increase in the number of teens who are vegetarians. Some choose this diet for moral or health reasons; others give little reason for their decision. The practice of vegetarianism varies greatly. Some individuals simply eliminate meat from their diet; others avoid all animal products, including dairy products. Vegetarians need to include cereals, legumes, and vegetables in their daily diet. These substances are needed to provide the essential amino acids necessary for growth and tissue repair.

Teens are likely to consume large amounts of soda or carbonated beverages. Some favor the regular colas with high caffeine and sugar contents. The diet colas are not any healthier because they contain artificial sweeteners, which have been the subject of much controversy.

SLEEP AND REST

The growth spurt that occurs during puberty and early adolescence causes an increased need for sleep. Adequate sleep and rest are needed to help maintain optimal health during this stage of development. Later in this stage the teen requires about 8 hours of sleep to be fully rested. There is a tendency for teens to stay up late to watch television or talk on the telephone. Staying up late causes them to be too tired to wake up in the morning or, when awakened, to be irritable. There also seems to a direct correlation between the lack of sleep and poor performance in school. Some teens are merely too tired during school to concentrate and learn.

EXERCISE AND LEISURE

Exercise is another important factor in helping teens maintain good health. It is one area in which many teens actively participate without much prodding. The teen's ability to perform skillfully and compete determines popularity and group acceptance. Some teens choose not to participate or select sporting activities.

Their reasons for avoiding these activities are either a lack of skill or a natural sedentary nature and personality. The activities that they may be interested in often challenge thinking rather than muscle coordination and skill. The patterns of exercise established during adolescence are likely to continue into the adult years.

Many teens take on part-time jobs in addition to school and extracurricular activities. These teens must learn to balance their time and other activities. Part-time employment offers the child spending money and exposure to the work world.

SAFETY

The leading cause of death in adolescence is accidents related to teens' increased motor abilities and strength, combined with their lack of judgment. This combination puts them at great risk for harm.

Although driving represents independence for the adolescent (Fig. 8–5), motor vehicle accidents are responsible for a majority of deaths during this developmental stage. Poor judgment, lack of driving skills, and failure to follow rules and use seat belts contribute to the high number of motor vehicle accidents involving teens.

Driver education courses and defensive driving programs can help reduce the number of accidents.

Teens must be encouraged not to drink and drive. Those who drink must be told to appoint a designated driver. Parents need to model proper behavior to help instill these values in their teens.

Sporting activities also account for many injuries during this stage. Teens often exercise little caution when competing in athletic activities. Proper physical exams must be done before a teen engages in any sporting event. Teens must adhere to the use of the proper protective equipment when participating in contact sports. It has been shown that protective clothing can further reduce the number of sporting

FIGURE 8–5
Driving is an important landmark for the adolescent, representing independence.

injuries. Boys tend to have more injuries related to contact sports such as football and hockey, whereas girls have more injuries relating to gymnastics. Some sport injuries have a seasonal pattern.

The incidence of injuries and deaths related to firearms is steadily on the increase for children and adolescents in this country. Many of these accidents could be prevented with proper storage and training in the use of firearms. An alarming number of the accidents involving firearms occur in or around the home. Parents need to take responsibility for the proper education and supervision of firearms. Some toy items have been found to cause injuries and lethal damage in the hands of children and teens. Many manufacturers have stopped producing toy guns that may be mistaken for real firearms and result in devastation and harm.

The second leading cause of deaths among all adolescents and young adults is related to homicides. Risk factors include race and socioeconomic status. Half of all homicides are associated with alcohol use. Both alcohol and drug use has further complicated vehicle-related injuries and deaths.

HEALTH PROMOTION

In general, teens' state of health is reflective of their habits and nutritional patterns. The number of acute illnesses decreases during this stage of development. Yearly medical checkups are suggested for this age group. This examination should include vision and hearing screening. Problems with eyesight should be promptly corrected with glasses or contact lenses. Teens must be instructed to avoid use of stereo headphones at high volume. Frequent exposure to excessively loud noise has been proven to lead to nerve damage and hearing loss. Some teens have even shown significant hearing loss as a result of frequent attendance at loud concerts. Weight and height measurements and nutritional guidance should be a part of each health visit. Dental examinations must be scheduled every 6 months or more frequently if there is dental decay or problems with malocclusion.

Blood pressure recordings must be monitored during the teen years to help detect any signs of abnormalities so that preventive measures and treatment may be promptly instituted. Blood cholesterol levels are examined and dietary interventions are offered to those individuals who appear to be candidates for high blood cholesterol. Girls are prone to anemia; therefore a complete blood count should be done at least yearly or more often. Symptoms such as fatigue, weakness, or excessive menstrual flow may be indicative of anemia.

Generally the teenager has a high resistance to early childhood illnesses. The well-functioning immune system still needs the support of booster immunization at ages 14 through 16 years for the prevention of diphtheria and tetanus. Proper nutrition and other healthy living practices will help determine the overall health of the individual.

As stated in Chapter 7, the teen should continue to be accessed for any signs of spinal abnormalities. Many teens need reminders about the importance of good posture to prevent musculoskeletal pain and deformity in the later years.

Teens also have many concerns and questions about skin care and hygiene. Teens are generally very sensitive about the condition of their skin, so much so that even the appearance of a small blemish will cause them to become distressed. They should be instructed in the basics of proper skin care. If they develop an acute case of acne, further medical treatment is necessary.

Depression

Depression, a prolonged feeling of sadness and unworthiness, is a serious problem affecting many teens. Stress from school, family, and personal relationships may overwhelm the teen and lead to this mood disorder. Teens are more prone to this condition because they spend more time in self-reflection, which may lead them to disappointment and despair. Signs of depression may go unnoticed by family and friends. The risk of suicide increases for the depressed. The common signs of depression are listed in Box 8–2.

BOX 8–2

Signs of Depression and Suicide

- Crying spells
- Insomnia
- Eating disorders
- Social isolation, withdrawal
- Acting-out behaviors: school phobias, underachievement, truancy, temper outbursts, substance abuse
- Feelings of hopelessness
- Unexplained physical symptoms
- Loss of interest in appearance
- Giving away things or possessions

Substance Abuse

Substance abuse refers to out-of-control use of tobacco, alcohol, and other drugs. Teenagers' need to be accepted makes them more likely to smoke. Even with the current legislation curtailing the sale of tobacco to minors, many easily obtain cigarettes. Antismoking campaigns have been instituted with minimal success. The need for early parental education and positive role modeling may be the best deterrent to the use of tobacco.

Experimentation with alcohol is considered to be another teenage rite of passage. Most teens begin drinking before the legal age. Drinking may occur at home, in school, or in other social settings. Some teens may feel that drinking helps them to deal with their feelings or avoid facing certain realities. Both tobacco and alcohol are used in the movies as a prop to portray sexuality. This greatly influences children and adolescents. Alcohol abuse occurs when the intake of alcohol interferes with day-to-day activities. Intense counseling and support groups may help the teen with any substance problem.

Drug experimentation can begin at any age but is most often seen during adolescence. Any drug, prescribed or nonprescribed, may be abused. Research indicates that the most common offender is marijuana. This starts with occasional use and may lead to habitual use. This behavior may then extend to the use of other drugs. Drug use has been shown to lead to other social problems, including sexual promiscuity, diseases, and pregnancy. Refer to Box 8–3 for signs of possible drug use.

BOX 8–3

Signs of Drug Abuse

- Altered sleep patterns: drowsiness, sleepiness, lethargy, or hyperactivity
- Mood swings
- Change in appetite
- Marked irritability
- Loss of interest in friends, school, and other activities
- Secretiveness
- Property or money losses
- Impaired judgment
- Change in hygiene or appearance

SUMMARY

1. Puberty or preadolescence is a period of rapid growth ending with reproductive maturity. In girls puberty ends with the onset of menarche; in boys puberty ends with the production of sperm.
2. The major changes associated with puberty include rapid physical growth, change in body proportions, and the development of primary and secondary sex characteristics.
3. Primary sex characteristics affect the growth and maturation of the gonads or sex glands.
4. The male gonads are the testes, which, when mature, produce sperm and the male sex hormone. The release of sperm, known as ejaculation, indicates functional maturity.
5. The female gonads are the ovaries. The production of ova and female hormones signals their maturity.
6. Menarche is the first menstrual period. Menstruation will occur monthly from puberty until menopause.
7. Secondary sex characteristics refer to all the changes that have no direct role in reproduction. These changes include hair growth, increased activity of the sweat glands, voice changes in boys, and widening of the hips and pelvis in girls.
8. Youths of this age tend to spend more time by themselves and move away from earlier friendships. In many settings teens are argumentative, causing their family relationships to change.
9. Adolescence is a transitional period that begins with sexually maturity and ends with physical maturity. This stage bridges the gap between dependence and independence, childhood and adulthood.

10. Dating begins around 15 to 16 years of age. It helps establish a means of social status, recognition, and recreation with the peer group.
11. Sex is given high priority at this time. The teen usually experiences a great deal of pressure to conform to the group's standards. Many teens engage in sexual intercourse during the adolescent period.
12. Sexual education must be provided in the home and further reinforced in school.
13. Masturbation is considered to be a normal part of sexual expression.
14. Sex education needs to be provided before the teenage period of development. During this period, teens continue to need advice about prevention of sexually transmitted diseases and pregnancy.
15. The major characteristics of adolescence include stormy emotions, feelings of insecurity, introspection, interest in experimentation and learning, and the testing of values and beliefs.
16. According to Erikson, the psychosocial task for this stage is the search for identity. Identity begins with the separation of the individual from the family. The movement away from the family expresses the teen's need for freedom and independence.
17. Adolescents are ambivalent about many issues, including loving and hating their family, wanting freedom and needing supervision, wanting to be a part of a group and wishing to be alone. Some of the commonly expressed emotions are anger, fear, worry, jealousy, envy, and happiness.
18. Peer relationships are very important at this stage. Peers share the same age, feelings, experiences, goals, and doubts in ways that parents can't.
19. Teens now forms cliques, groups, and gangs. They have a strong need to be accepted by the members of these groups. Group acceptance helps them to feel happy and confident; nonacceptance leads to feelings of alienation and resentment.
20. Society places many demand on teenagers. They are expected to select a vocation and think about their future.
21. Maturation of the central nervous system causes a shift from concrete thinking to formal operational thought processes. This thinking style is more logical. Teens can think abstractly and reason scientifically. School is at the center of activity for the teenager. Success in school depends largely on socioeconomic background, family relationships, peer influence, and social pressures.
22. Moral judgment is based on the learned principles of right and wrong.
23. Adolescents also develop a spiritual awareness. They question, philosophize, and compare religions.
24. Discipline during adolescence is very important. Many of the conflicts between parents and teens are based on the choice of friends and issues of dating.
25. The rapid growth that occurs during adolescence calls for an increase in nutritional requirements. Eating habits change during adolescence; meats and potatoes are favored over fruits and vegetables. Snacks are chosen for their accessibility and taste.
26. In early adolescence teens have an increased need for additional sleep. About 8 hours are needed for teens to be fully rested. Staying

up late causes the teen to be tired and irritable in the morning. Lack of enough sleep appears to relate to poor school performance.
27. Exercise is important to help maintain the teen's state of health.
28. The leading cause of death during adolescence is accidents.
29. Teens' general state of health is reflective of their habits and nutritional practices. Teens require yearly medical checkups. Proper nutrition and other healthy practices will help contribute to an overall healthy lifestyle.
30. The incidence of depression is higher than at earlier stages of development because of the increased stress and demands of this stage. Any indication of prolonged sadness or depressed mood must be carefully assessed and monitored.
31. Suicide prevention is an important safety measure to be familiar with when dealing with teenagers.
32. Experimentation with alcohol, drugs, and tobacco is common during this stage of development. Teens often engage in these unhealthy practices to gain acceptance from their peers.

CRITICAL THINKING Anna Avery, 14 years of age, arrives at the walk-in clinic seeking advice. She tells the healthcare provider that she has had sexual intercourse once with her boyfriend. She admits that she will continue to be sexually active. She expresses a concern about not wanting to become pregnant or infected with a sexually transmitted disease. Anna feels that she cannot discuss these concerns with her parents. Outline a simple teaching plan to present to Anna.

Multiple-Choice Questions

1. The major psychosocial task of adolescence, according to Erikson, is:
   a. Autonomy
   b. Identity
   c. Intimacy
   d. Trust

2. Menarche is best defined as:
   a. The development of body hair
   b. The onset and release of female hormones
   c. First menstrual flow
   d. The development of breasts

3. Which of the following physical changes usually occur during adolescence?
   a. A decline in male hormone production
   b. A decrease in blood volume
   c. A decrease in sebaceous secretions
   d. An increase in muscle strength and endurance

4. Bulimia is characterized by:
   a. Periods of starvation
   b. Periods of bingeing
   c. Gradual increases in weight
   d. Gradual loss of bone mass

5. Moral development in the beginning of adolescence is demonstrated by:
   a. The acceptance of society's rules and standards
   b. The questioning of existing rules and standards
   c. The teen being more self-centered
   d. A strong individual moral code

6. One of the objectives fulfilled by dating is:
   a. To establish adult behaviors
   b. To reinforce principles of justice
   c. To fulfill personal and social status
   d. To help promote independence

Suggested Readings

Ahmann, E, and Bond, NJ: Promoting normal development in school-age children and adolescents who are technology-dependent: A family-centered model. Pediatr Nurs 18(4):399–405, 1992.
Bergmann, PE, Smith, MB, and Hoffmann, NG: Adolescent treatment: Implications for assessment, practice guidelines and outcome management. Pediatr Clin North Am 42(2):453–472, 1995.
Christophersen, ER: Discipline. Pediatr Clin North Am 39(3): 395–411, 1992.
Delaney, CH: Rites of passage in adolescence. Adolescence. 31(121)893–897, 1996.
Emmons, L: 1996 The relationships of dieting to weight in adolescents. Adolescence 31(121):167–177, 1996.
Epps, RP, Manley, MW, and Glynn, TJ: Tobacco use among adolescents: Strategies for prevention. Pediatr Clin North Am 42(2):389–399, 1995.
Finney, J, and Weist, MD: Behavioral assessments of children and adolescents. Pediatr Clin North Am 38(3):369–378, 1992.
Fuller, PG, and Cavanaugh, RM: Basic assessment and screening for substance abuse in the pediatrician's office. Pediatr Clin North Am 42(2):295–307, 1995.
Gidding, SS: Relationships between blood pressure and lipids in childhood. Pediatr Clin North Am 40(1):41–49, 1993.
Grossman, DC, and Rivara, FP: Injury control in childhood. Pediatr Clin North Am 39(3):471–485, 1992.
Hendee, WR: The Health of Adolescents: Understanding and Facilitating Biological, Behavioral, and Social Development. Jossey-Bass, San Francisco, 1991.
Hockenberry-Eaton, M, Richman, MJ, Dilorio, C, Rivero, T, and Maibach, E: Mothers' and adolescents' knowledge of sexual development: The effects of gender, age, and sexual experience. Adolescence 31(121):35–47, 1996.
Kelly, DP, and Aylward, GP: Attention deficits in school-aged children and adolescents: Current issues and practice. Pediatr Clin North Am 39(3):487–512, 1992.
Kershner, R: Adolescent attitudes about rape. Adolescence 31(121):29–33, 1996.
Kidwell, JS, Dunham, RM, Bacho, RA, Pastorino, E, and Portes, P: Adolescent identity exploration: A test of Erikson's theory of transitional crisis. Adolescence 30(120):785–793, 1995.
Luer, RM, Clarke, WR, Mahoney, LT, and Witt, J: Childhood predictors for high adult blood pressure: The Muscatine study. Pediatr Clin North Am 40(1):23–40, 1993.
McGory, A: Education for the menarche. Pediatr Nurs 21(5):439–443, 1995.

Meeus, W, and Dekovic, M: Identity development, parental and peer support in adolescence: Results of a national Dutch survey. Adolescence 30(120):931–943, 1995.
Montgomery, JW: Easily overlooked language disabilities during childhood and adolescence: A cognitive-linguistic perspective. Pediatr Clin North Am 39(3):513–524, 1992.
Morrison, SF, Roger, PD, and Thomas, MH: Alcohol and adolescence. Pediatr Clin North Am 42(2):371–387, 1995.
O'Malley, PM, Johnston, LD, and Bachman, JG: Adolescent substance use: Epidemiology and implications for public policy. Pediatr Clin North Am 42(2):241–260, 1995.
Patton, D, Kolasa, K, West, S, and Irons, TG: Sexual abstinence counseling of adolescents by physicians. Adolescence 30(120):963–969, 1995.
Patton, LH: Adolescent substance abuse: Risk factors and protective factors. Pediatr Clin North Am 42(2):283–291, 1995.
Rocchini, AP: Adolescent obesity and hypertension. Pediatr Clin North Am 40(1):81–92, 1993.
Rogers, PD, Sperano, SR, and Ozbek, I: The assessment of the identified substance abusing adolescent. Pediatr Clin North Am 42(2):351–370, 1995.
Werner, MJ: Principles of brief intervention for adolescent alcohol, tobacco, and other drug use. Pediatr Clin North Am 42(2):335–349, 1995.
Whaley, L, and Wong, D: Nursing Care of Infants and Children. Mosby-Year Book, St. Louis, 1991.
White, FA: Family processes as predictors of adolescents' preferences for ascribed sources of moral authority: A proposed model. Adolescence 31(121):133–143, 1996.

Chapter 9

Chapter Outline

Early Adulthood

Key Words

aerobic exercise
basal metabolic rate
carcinogens
cholesterol
compatibility
free radicals
gingivitis
hypertension
insomnia
intimacy
introspection
mammography
Mantoux skin test
obesity
occult blood
osteoporosis
Papanicolaou test
presbyopia
proximity
reaction time
reciprocity
resistance exercise
respectability
saturated fats
sexuality
sun protection factor (SPF)
unsaturated fats
vital capacity

Learning Objectives

At the end of this chapter, you should be able to:

- List four goals for the early adult period of development.
- Describe three physiological changes that occur during early adulthood.
- Describe the psychosocial task as identified by Erikson for the early adulthood period.
- Name three nutritional concerns for young adults.
- Describe two health screening tests important for women in the early adult period of development.

Early adulthood covers the period from age 20 through the early 40s. This stage of development is generally described as a stable time of growth. Gradual biological and social changes are expected at this stage. As some body systems grow and develop, others begin to show the effects of aging. All early events, experiences, and patterns of growth help shape and prepare individuals for their adulthood.

Adulthood is a period that most adolescents have anticipated and striven for. Entrance into this stage is usually accompanied by positive feelings, dreams, and aspirations. The goals for this time include choosing and establishing a career, fulfilling sexual needs, establishing a home and family, expanding social circles, and developing maturity. Toward the completion of this stage, adults begin to compare their early dreams with their accomplishments. As this occurs, they must reconcile the differences and accept the reality or institute changes.

PHYSICAL CHARACTERISTICS

Height and Weight

Physical growth is completed in adulthood. Men continue to show growth in the vertebrae until age 30. This growth adds about 3 to 5 mm in height. Women usually attain their full stature before their twenties.

Bone and Muscle Development

Peak bone mass is attained by age 35. There is a gradual loss of bone mass after this age in women. Although bone growth stops, bone cells are replaced at the site of any injury. Exercise helps to increase endurance, strength, and muscle tone. Actual muscle mass differs in men and women based on nutrition, exercise, and amounts of the hormone testosterone; for this reason, men usually have more muscle mass than women. Increase in muscle mass is not dependent on an increase in muscle cells. Muscle capacity for sports varies with age. The ability to engage in vigorous sports such as tennis and football declines after the first half of adulthood, while interest in other sports like golf may first begin in the later part of adulthood. Capacity and maximum work rate without fatigue begin to decline after age 35. Injury occurring during this stage best responds to rest and immobilization.

Dentition

Wisdom teeth erupt toward the end of adolescence or in the early 20s (see Chapter 8). Failure of these teeth to erupt may lead to pain and overcrowding and may require surgery. Gum disease, or **gingivitis**, affects many adults and is considered preventable. It is the major cause of tooth loss in the adult years. The need for proper care of the teeth and gums cannot be overemphasized. Proper care includes regular brushing, flossing, and avoidance of excessive sweets. Good oral health includes visits to the dentist every 6 months.

Development of Other Body Systems

All organs and body systems are fully developed and matured by this age. Changes in body shape, growth of body hair, and muscle development slowly continue through the 20s.

Maximum cardiac output is reached between ages 20 and 30; after that cardiac output gradually declines. During the adult years the heart muscles thicken, with fat deposits in the blood vessels producing a decrease in blood flow. Certain practices, including intake of alcohol, tobacco, and foods high in cholesterol content, may increase the individual's risk for cardiovascular disease. The heart and vessels become less elastic with advancing age. This rigidity may contribute to the decreased cardiac output and the increased blood pressure commonly seen in later adulthood.

The ability of the lungs to move air in and out is known as **vital capacity**. Vital capacity decreases between ages 20 and 40. Peak respiratory function for men is at age 25 and for women at age 20. This gradually declines owing to loss of elasticity in the lung. Adults who smoke tend to lose elasticity more rapidly than nonsmokers. This loss of elasticity also leaves the individual more susceptible to respiratory infections. Exercise can maintain and maximize lung capacity.

Appetite remains unchanged in this stage. After age 30 the gastric secretions and digestive juices diminish significantly. Poor eating habits lead to common gastric discomforts and indigestion. The **basal metabolic rate**, which is the amount of energy that an individual uses at rest, decreases with advancing age. This change may result in an increase in weight even when dietary habits remain unchanged. Adults need to try to maintain normal bowel elimination with a diet containing roughage and adequate fluids. A balance of diet and exercise, along with regular patterns of elimination, promotes normal bowel functioning. Individuals should report to their physician any change in their normal patterns of elimination so that that proper medical investigation and treatment can be instituted.

Skin cells undergo some changes as a result of exposure to the sun and pollutants in the environment. Excessive exposure to ultraviolet rays may produce cancer of the skin, particularly in light-skinned individuals. (See "Exposure to Carcinogens" under section on health promotion for further information relating to skin protection.) Adolescent acne usually clears up by adulthood. For the few cases that do not, a number of different treatment regimens can be offered.

Both the number of cells in the nervous system and the size of the brain decrease after puberty. Changes in sensation and perception can be recognized during this stage. However, speed and accuracy of these perceptions are not yet affected. **Reaction time**, the speed at which a person responds to a stimulus, increases noticeably between ages 20 and 30. Visual acuity may decline after age 25 owing to decreased elasticity and increased opacity of the lens. By age 40 there is often a decreased ability to see objects at close distance. This condition, known as **presbyopia**, advances with age. Corrective lenses can correct vision in the person suffering from this condition. Hearing ability is best at age 20; after that there is a gradual loss of hearing for high-frequency tones. Hearing loss occurring at this point usually has little effect on the individual's activities of daily living. Excessive exposure to loud noise from music or work may accelerate hearing loss. Most adults learn to compensate for the minor losses and in so doing can maximize their performance.

The body system that is actually functioning at peak capacity is the reproductive system. In women, the menstrual cycle is well established. Women should report irregular patterns of menstruation or serious discomfort to their physician. Generally, men are free of reproductive problems at this stage. One concern that may threaten the couple's sexuality and emotional well-being is infertility. Couples who are experiencing difficulty conceiving should seek counseling and medical supervision.

VITAL SIGNS

The normal resting heart rate for the adult ranges from 60 to 90 beats per minute. The normal respiratory rate ranges from 18 to 24 breaths per minute. Normal blood pressure readings range from 110 to 130 systolic and 70 to 90 diastolic.

DEVELOPMENTAL MILESTONES

The major developmental milestones for this age group include choosing and establishing a career, fulfilling sexual needs, establishing a home and family, expanding social circles, and developing maturity.

Motor Development

Most individuals have reached peak physical efficiency during this period of development. Muscle strength and coordination peak in the 20s and 30s and then decline gradually between ages 30 and 60. The areas of greatest strength include the back, arms, and legs.

Sexual Development

Adults must first come to terms with themselves as sexual beings and then become comfortable with their own sexuality. **Sexuality** is a broad term that includes anatomy, gender roles, relationships, and thoughts, feelings and attitudes about sex. Many factors influence the development of sexuality: biological development, personality traits, cultural and social influences, and religious and ethical values. Education expands the adult's knowledge and understanding of sexual behavior, permitting the development of positive feelings. This helps further communication and openness in intimate relationships. The goal is to enable people to achieve pleasure and sexual satisfaction in their relationships.

Part of fulfilling sexual needs is the adult's ability to experience and share love. Romantic love is a deep emotional experience, with mutual sharing of warm and tender feelings. Unlike individuals at earlier stages, the mature adult now has the basis for establishing this intense relationship. Romantic love incorporates intimacy and passion. All love is reciprocal and allows for giving and sharing with one another. The mutuality of sharing and the bonds of commitment foster a sense of security between the individuals. Love brings people together and is more than just a sexual experience.

For most men and women, sexual concerns are usually stable during this stage of development. Most adults, by their mid-20s, have already established comfortable patterns of sexual behavior, and most feel comfortable with their masculinity or femininity.

Many studies have been done on human sexual response. Although feelings and attitudes vary greatly among individuals, basic responses to sexual arousal have common features. The best-known study was by Masters and Johnson, who described the cycle of human sexual response by dividing the response into four distinct stages: excitement, plateau, orgasm, and resolution. The *excitement phase* begins with feelings and sensations that produce muscle tension and vasocongestion in the reproductive organs. A state of heightened excitement occurs during the *plateau phase*, which occurs just before orgasm. During the *orgasmic phase* there are rhythmic contractions in the vagina and penis and, in the male, ejaculation (release of semen). The other physiological responses to sexual arousal include increases in blood pressure, respiration, heart rate, and muscle tension; and engorgement or swelling of the genital tissues. These responses add to the overall arousal state. During the *resolution phase*, the reproductive organs return to their unaroused state. Men have a brief refractory period during which they cannot have a repeat orgasm. It is possible for women, if stimulated and desirous, to have repeated orgasms, one following another. Most recent research in the area of sexual behavior focuses on the importance of integrating the mind and body to achieve a satisfying, healthy sexual experience.

Psychosocial Development

By the time adults reach their 20s, they need to have developed a strong sense of identity. This ego identity or sense of who they are will allow them to accomplish the next task, as described by Erikson: **intimacy**. Erikson broadly described intimacy as not only sexual intimacy, but emotional intimacy between lovers, between parents and children, and between friends. This definition of intimacy involves warmth, love, and affection. Individuals must be capable of giving of themselves in an emotional relationship. **Introspection**, or self-reflection, is the tool that is needed to permit the sharing of innermost thoughts. The individual must learn to be truly open and capable of trust. Intimacy is not just sex, but who we are and how we express ourselves in our male or female roles. These roles are affected by culture and time. A radical change has taken place in the female role since the early 19th century. Today's women may share with a partner or manage alone the many responsibilities in the home, workplace, and community. Similarly, the male role has changed to encompass not only breadwinner but, in some cases, homemaker and caregiver.

Adults who are uncertain of their identity often shy away from meaningful relationships and enter casual interactions that lack interrelatedness. This may lead to isolation and self-absorption. Without trust and commitment, these relationships are usually unfulfilled and doomed to failure.

Choosing and Establishing a Career

It is necessary to understand work roles and their meaning in order to better understand adult life. Certain events such as being hired, promoted, fired, and retired are considered critical milestones in the work cycle of the adult. Work roles form the basis of one of the major social roles of adulthood. Most adults work. Work makes possible personal, social, cultural, and financial survival. Work roles affect the individual's sense of identity because in our society people are often judged by what they do for a living and how much they earn. Work has different meanings for different people. For some, it represents prestige and social recognition; for others, it is a source of disappointment. Work may enhance self-worth, respect, and creativity. Lastly, work may represent service to others.

Both men and women enter the workforce with hopes of upward mobility, that is, a better job, an increase in salary, or a promotion (Fig. 9–1). Wages, promotions, and the ability to accumulate expensive possessions are used as measurements of work role success. The experience for women in the workforce may be different from that of men. Women are faced with pressures of family, self, and work. The demands of an occupation may have to be balanced with the demands of marriage and childbearing. These conflicts may lead to career obstacles and undue stress. For some women, work allows economic independence and may create less pressure to marry. Many choose marriage, but for other reasons than economic security. Women's occupations are also changing. In a recent study, college women indicated that they were pursing careers in law, business, medicine, and engineering. These careers are no longer male-dominated. Because of these changes, men now need to adjust to the new role of women in the workforce and as family providers. Both internal and external pressure is placed on all individuals to succeed in their occupations.

Women in the workplace encounter sexual harassment or inequities more frequently than men. Sexism, like other prejudicial stereotyping, has an adverse effect on society. Sexism refers to all the attitudes, beliefs, laws, and actions that discriminate on the basis of gender. We see sexism today evidenced by stereotyping and unequal treatment of individuals. One consequence of negative stereotyping is that victims may tend to believe that the portrayal is true; as a result they may undervalue or further degrade themselves. Sexism is still evident in the employment status of women. For example, there is still a significant earnings gap between men and women: in 1992 pay for women was 75.4 percent that of men. There have been attempts to make the public more aware of these issues and make workers and management more sensitive to people's rights and feelings. See Box 9–1 for signs of sexual harassment.

FIGURE 9–1
Young adults place great importance on work and job success.

BOX 9–1

Signs of Sexual Harassment

- Continual or repeated verbal abuse of a sexual nature
- Graphic sexual comments, gestures, or postures
- Display of sexually suggestive objects
- Sexual propositions, threats, or insinuations suggesting that if the person refuses to submit sexually, his or her employment, wages, or status will be adversely affected

Periods of unwanted unemployment create increased stress for individuals, their families, and their support systems. Prolonged joblessness can cause serious psychological and social problems. Whether it is permanent or temporary job loss through downsizing, restructuring, or otherwise, the individual suffers from a loss of steady income and, often, a loss of self-worth as well. Lengthy unemployment may eventually lead to depression and social isolation.

Establishing a Home and Family

For many, early adulthood is the time to establish a home and family. Many young adults choose to leave their family of origin and start a home of their own after their adolescent years. Finding a place to live and call their own is an important step for young adults. Where they settle down largely depends upon available jobs and income. Some young adults want to remain close to their family of origin. A major decision for the adult is whether to choose a mate or to remain single.

Some adults decide to marry, whereas others choose other types of relationships. Their relationships may be long-lasting or short-lived. This may depend upon the individual's own goals and needs (Fig. 9–2). Adults who have not resolved the conflict of identity usually experience the most difficulty in their close relationships. All adults involved in a relationship must establish clearly defined roles to

FIGURE 9–2
Young adults seek meaningful, loving relationships.

minimize conflicts. The decision to start a family and raise children is a very individual choice. Some adults become involved in their careers and delay parenting. Others may make the choice to remain childless. Still others choose to raise children as a single parent. Chapter 2 has additional readings on family styles and arrangements.

Expanding Social Circles

Adults tend to select friends on the basis of similarity of life stage, such as age of children, duration of marriage, occupational status, or community interest. Adult friendships often last over long periods of time and survive periods of separation. Young adults share feelings, experiences, and confidences with their friends. Friendships may be either acquaintances or intimate relationships. Characteristics of intimate friendships include reciprocity, compatibility, respectability, and proximity. **Reciprocity** refers to mutual helping and supporting interactions between friends that allow them the freedom to rely on one another. The central theme of reciprocity is giving and receiving. **Compatibility** describes the feeling tone of the relationships. The components of compatibility are comfort, ease of the relationship, and friendship. **Respectability** emphasizes role modeling and valuing. **Proximity** describes the frequency of interaction and the duration of the relationship; these two factors are more important than geographic location. Adult friendships occur in a variety of settings: in the home, at work, and in the community. These relationships are necessary as they provide individuals with emotional support and stability.

Developing Maturity

Mature adults have developed both an internal and external system of controls and restraints. This allows them to behave in an acceptable manner. Mature adults have established a philosophy of life that incorporates their beliefs and ethical values. This helps them make decisions and choices and maintain their sense of individuality. Adults have a broad perspective and are open to suggestion but not overly influenced by others. They are capable of living, sharing, caring, and respecting others. Another sign of maturity is the individual's ability to develop an interest in the community's needs. Mature adults are able to take responsibility for their actions. They are able to deal with problems or setbacks without losing site of their goals.

Cognitive Development

Unlike persons in earlier stages of development, the adult is no longer primarily egocentric. Adults are therefore capable of being objective and of looking at issues from a wider perspective. Cognitive ability draws on the individual's ability to solve problems and use information. It determines the how and why of knowledge. Most adults are at the level of formal operational cognitive functioning. This permits them to attain an increased amount of learning or function at their peak intellectual level. Injury or insult to physical or emotional health may have adverse effects on cognitive development and learning. By drawing on their past experiences, adults have an increased ability to reason, solve problems, and set priorities.

Intelligence is measured by the content of what is known. Most intelligence is measured by testing. Tests usually ask for the recall of a body of knowledge acquired during schooling. People from lower socioeconomic levels may have lower intelligence scores; research has shown that they would score higher if they were given the same learning opportunities as people in higher socioeconomic levels.

Approximately 35 percent of young adults attend college or vocational school. School helps adults to organize their time, expand their awareness, and sharpen their understanding of the world.

Some older adults return to school after many years and find that it may take them a little longer to adjust to the learning environment. One adult learning theorist suggested that the best climate for adult learners is one of mutual respect, trust, support, and caring. Adults learn at different rates because of individual differences. They usually have more than one reason for learning. They are motivated by things that have personal meaning and importance to them. Reinforcement is the force that helps them continue their learning process. Positive reinforcers for the adult learner include praise, social approval, and recognition. These positive reinforcers are stronger motivators than coercion and force.

Moral Development

Most adults are in the postconventional stage of moral development. They have the capacity to choose the principles and rules by which they live. For many individuals moral issues are not a matter of absolute right or wrong but need to be viewed in the context in which they occur. For example, adults may know that killing is wrong, yet during war they are able to report to active duty and perform the duties of a soldier and kill if necessary. Under these circumstances this action would be considered an honorable and moral act.

Highly moral individuals respect the rights of others. Morality is not just a rule but a code of behavior to guide one's actions. Some views on morality extend beyond love, ethics, and justice to a state in which one finds mutual satisfaction. This leads to true understanding of one's self and others. This interpretation of moral development is sometimes described as a feminist perspective of morality. As with other developmental issues, morality is a highly individual matter.

Many adults exhibit a strong interest in religion, sometimes returning to the religious teachings of their own upbringing in order to teach religion to their offspring. In most families the mother's religious values and beliefs are more likely to be practiced than the beliefs of the father.

NUTRITION

A sound diet is crucial to a person's general state of health at any age. Dietary needs in the adult years differ little from those in adolescence. Caloric requirements are based on the adult's age, body size, amount of physical activity, and gender. Men generally need between 2700 and 3000 calories per day, whereas women need only 1600 to 2100 calories per day. Each individual needs to adjust his or her calorie intake based on lifestyle (active versus sedentary) to help maintain desired body weight.

It is recommended that 15 percent of adults' daily intake be in the form of protein. Protein sources include dairy products, meat and fish, legumes, soy products, and nuts. Recent research has shown that adults who consume fish as a part of their diet are at lower risk for heart disease. Certain types of fish, such as salmon, trout, mackerel, and bluefish are especially recommended because they contain omega-3 fatty acids, which help to lower the total serum cholesterol levels. Currently, scientists are reluctant to recommend fish oil supplements because their long-term effects are not known. For vegetarians, flaxseed oil may be used as a supplemental source of omega-3 fatty acids.

Only a very small amount of fats are needed in the diet to maintain good health. Extra fats only serve to add additional calories and contribute to obesity. A diet high in fat also raises blood cholesterol levels. **Cholesterol** is a component of many foods in our diet. The liver manufactures and filters out excess cholesterol. Cholesterol is an essential component of cells in the brain, nerves, blood, and hormones. However, an increase in the serum cholesterol is considered the major cause of coronary artery disease. People with cholesterol levels under 200 mg are at least risk for coronary artery disease, while those with levels over 240 mg are at greatest risk. To maintain healthy cholesterol levels, adults require only about 30 percent of their total caloric intake from fats. The American Heart Association recommends that women eat no more than 6 oz of meat per day and men eat no more than 7 oz of meat per day.

Foods high in saturated fats should be kept to a minimum. **Saturated fats**, which become solid at room temperature, are found in meat, poultry, and dairy products (butter, cream, whole milk), as well as in palm oil and cocoa butter. Different cuts of meats vary in their saturated-fat content. Those meats with visible fat usually have a higher saturated-fat content. Trimming the visible fat from the meats and removing the skin from poultry can help reduce the total saturated-fat content. Eggs and organ meats (liver, heart, and kidney) are very high in saturated fat and therefore should be used sparingly. Baking and broiling are preferable to frying and sautéing, as they render the fat from meat without adding extra oil.

Unsaturated fats are likely to be liquid at room temperatures. These fats are derived from plant sources such as corn, cottonseed, safflower, and soybeans. The terms *monounsaturated* and *polyunsaturated* refer to the compound's specific chemical composition. Recent research indicates that a diet low in saturated fat and high in monounsaturated fat decreases the risk of colon and rectal cancers. Refer to Table 9–1 for summary of different types of fats.

TABLE 9–1
TYPE OF FAT IN DIFFERENT FOODS

Saturated fats (high in cholesterol)	Liver, kidneys Eggs Shrimp, lobster, oysters Coconut and palm oils Whole milk, butter, cheese Red meat
Monounsaturated fats	Canola, olive, and peanut oil; avocados; olives; almonds, cashews, and filberts
Polyunsaturated fats	Corn, cottonseed, safflower, sunflower, sesame, and soybean oils
Omega-3 fatty acids	Halibut, mackerel, herring, salmon, sardines, fresh tuna, trout, and whitefish; flaxseed oil

The daily caloric intake for adults should contain 50 to 60 percent carbohydrates. Complex carbohydrates, such as grains (wheat, rice, corn, and oats), peas and beans, and starchy vegetables (potatoes and yams), are rich in vitamins and minerals and high in fiber content. Fiber promotes bowel elimination. Daily vitamin supplements can be taken but should never be used as a substitute for natural food sources or taken in therapeutic doses unless prescribed by a physician.

The young adult must safeguard against rapid bone loss and the development of osteoporosis. **Osteoporosis** is a disorder characterized by decreased bone mass resulting from the loss of minerals from the bones. This disorder primarily affects women beginning in the fourth decade. There are two main reasons for the high incidence of osteoporosis: (1) women have proportionately less bone mass than men; (2) as menopause approaches, women's estrogen levels decline, causing the rate of mineral resorption to exceed the rate of bone formation. Adequate calcium intake, regular exercise, and hormone replacement therapy may help decrease the risk of osteoporosis.

Calcium and vitamin D are essential for the maintenance of strong bones and teeth. Most women consume far less calcium than the recommended 1000 to 1500 mg per day. Good sources of calcium and vitamin D include milk and dairy products, meats, dark green vegetables, canned salmon, sardines, and tofu.

Free radicals are chemical substances produced during metabolism; it is suspected that they play a role in cellular aging. Vitamins C and E have been identified as antioxidants, or substances that can interfere with the formation of free radicals. Vitamin E can be found in vegetable oils, wheat germ, nuts, legumes, and green leafy vegetables. Vitamin C is found in fruits and vegetables. Vitamin C is not stored in the body and must therefore be supplied daily.

Hypertension, or high blood pressure, is a condition that places the individual at greater risk for heart disease and stroke. It affects many adults. For reasons that are not understood, African Americans have a higher incidence of hypertension than other ethnic groups. Some studies indicate that foods high in sodium may cause elevated blood pressure. Individuals with a history of hypertension should limit or avoid excessive intake of sodium-rich foods. Sodium is found in many prepared foods, including prepared or cured meats and fish, soups, sauces, condiments, and certain snack foods.

SLEEP AND REST

Adults need an average of 7 to 9 hours of sleep each night. Adequate sleep helps the adult function with maximum productivity. Some individuals may complain of **insomnia**, or inability to sleep. Manifestations of insomnia include taking a long time to fall asleep, awakening frequently during the night or too early in the morning, and feeling tired and unrested on awakening. Diet, stress, fatigue, and poor physical health may be contributing factors. Sleep difficulties are sometimes a side effect of medication. The excessive use of caffeine, alcohol, nicotine, sleeping pills, and other drugs can further disturb the body's natural sleeping patterns. Insomnia that persists beyond a couple of weeks may indicate a medical problem that warrants further attention.

The following measures may help promote better sleep:

- Avoid large meals before bedtime.
- Plan regular exercise in the early afternoon.
- Follow a bedtime routine.

- Practice relaxation before bedtime.
- Use the bed only for sleep, not for reading or watching TV.
- Establish a schedule to awaken each day at about the same time.

EXERCISE AND LEISURE

Physical fitness can improve at any age with regular participation in exercise (Fig. 9–3). **Aerobic exercises** work the large body muscles, elevating cardiac output and metabolic rate. Aerobic exercises help to develop muscle fitness, endurance, power, and flexibility. Aerobics are the best form of exercise for burning calories. Brisk walking, cycling, and running are some examples of aerobic exercises. **Resistance exercise**, such as weight lifting, burns fewer calories but builds muscle mass and maintains metabolic rate.

To improve cardiovascular health, it is recommended that the adult exercise three to five times a week for about 20 minutes at each session. After several weeks of training, the person will have achieved maximum cardiac output, thereby increasing speed of oxygen delivery to the tissues. The lack of proper exercise can produce fatigue, headache, backache, and complaints of joint pain. Exercise should be incorporated into the adult's daily routine. Many social experiences may be built around the adult's exercise program (Fig 9–4).

SAFETY

Safety concerns for the adult are similar to the concerns discussed in Chapter 8. Adults now need to expand their safety concerns beyond themselves to that of their children and other family members. Safety in the home is a topic that must always be emphasized and practiced. Fire safety and prevention in the home must be addressed, including the use of extinguishers and smoke detectors and proper storage of flammable materials. Batteries in smoke detectors should be changed twice a year to ensure proper functioning. Each family member must be aware of a plan for

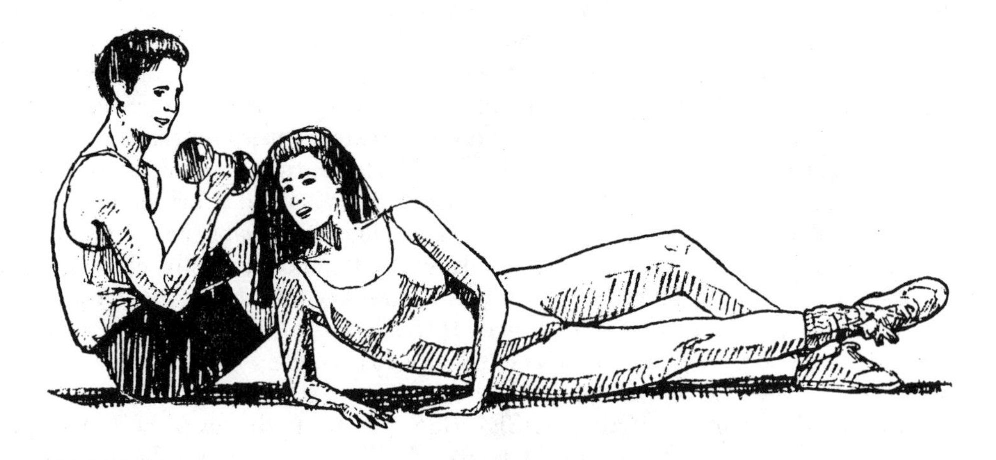

FIGURE 9–3
Young adults like to engage in exercise.

FIGURE 9–4
Pleasurable leisure activities help decrease stress.

escape in the event of fire in the home. Adults can best teach safety measures by setting good examples for their children to observe and follow.

HEALTH PROMOTION

Health assessment during the adult years should consist of a yearly physical examination. As a part of the examination a complete blood analysis should be performed so that any early problems or abnormalities can be identified and corrected. Because of the increase in tuberculosis cases in the United States, everyone should receive screening for TB in the form of a **Mantoux skin test**. A follow-up chest x-ray must be done if the results are positive. Electrocardiogram (ECG) testing is useful in providing a baseline cardiac picture. Blood pressure screening and weight assessment must be part of the adult's yearly health assessment. Early detection of health problems can lead to prompt intervention and ultimately protect against future illness.

Gynecological concerns include problems with conception, infertility, and menstrual discomfort or disorders. The **Papanicolaou test** is used to screen for cancer of the cervix. There are five levels of test results: class 1, the absence of abnormal cells; class 2, atypical but nonmalignant cells; class 3, abnormal cells; class 4, cells that are possibly but not definitively malignant; and class 5, conclusive for cancer. All women should have a yearly Pap smear. If there is a familial history of cervical cancer, this test should be instituted as early as adolescence.

All women over the age of 20 should be familiar with the correct method for performing breast self-examination (BSE). BSE has been shown to be the single most important examination used to detect early disease. It should be performed once a month about 1 week (7 days) after the end of the menstrual period. Box 9–2 outlines the steps in breast self-examination. The test begins in front of the mirror

BOX 9–2

Breast Self-Examination (Monthly)

While standing in front of the mirror:

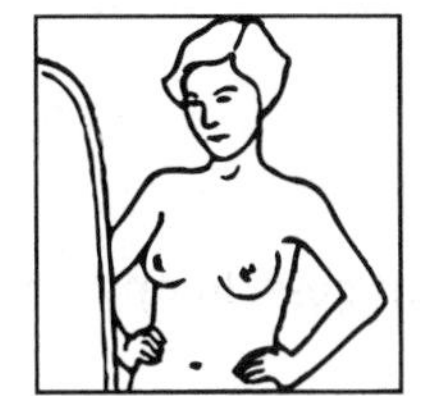

1. Keep your arms at your side and then raise them above your head.
2. Look carefully at the size, shape, and contour of each breast.
3. Look for puckering, dimpling, or changes in the skin texture.
4. Note if there is any nipple discharge.

While lying down on the right side:

1. With a pillow under the right shoulder, place your right hand behind your head.
2. With the fingers of your left hand, press gently in a circular motion, starting at the outside and spiraling toward the nipple.
3. Examine your underarm and the area below your breast.
4. Repeat for your left breast.

While standing in the shower:

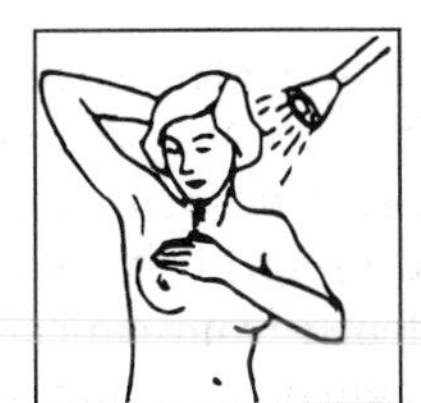

1. Raise your right arm and use your left hand to examine your right breast.
2. Using a circular motion, start from the outer area and proceed inward.
3. Gently feel for any lump or thickening.
4. Repeat for left breast.

with an inspection of the breasts for gross irregularities. Palpation for lumps or irregularities can then be done standing up, for example, in the shower, or in a supine position, for example, while lying in bed. Breast cancer can occur in men as well as women, although this is rare; therefore, men as well as women should examine their breasts and report any unusual lumps or growths to their physicians. **Mammography**, or breast x-ray, should be initiated at age 40 and performed every other year until age 50 and yearly thereafter. If the client has a family history of breast disease, a yearly mammogram is recommended.

In men, health screenings should include monthly examination of the testicles for early detection of tumors or other growths. Box 9–3 describes testicular self-examination.

All adults, men and women, should have an annual rectal examination that includes a simple test for **occult blood** (hidden blood) in the stool. The presence of occult blood may indicate any one of several gastrointestinal diseases.

Adults should have a tetanus booster every 10 years. Additional immunizations may be needed for those who are traveling outside the country. This kind of information may be obtained at the physician's office or at the local health department.

BOX 9–3

Testicular Self-Examination (Monthly)

1. Stand in front of the mirror.
2. Look at the appearance of the scrotum.
3. Examine each testicle using both hands.
4. Rotate between the thumbs and forefingers.
5. Report any dull pain in the groin, change in appearance, firmness, lumps, or irregularity.

The state of health and life practices may contribute to the development of heart disease. The risk factors that contribute to heart disease include lack of physical exercise, smoking, and elevated blood cholesterol and blood pressure levels. To control these risk factors, individuals must engage in moderate activity, avoid cigarette smoking, manage weight through appropriate diet, and comply with their prescribed medication regimen. Recent statistics have shown a decline in the number of deaths from heart disease due to the population's healthier lifestyle practices.

Exposure to Carcinogens

The National Cancer Institute estimates that about 80 percent of all cancer cases are related to lifestyle practices. Many cancers can be prevented by avoiding **carcinogens** (cancer-producing agents) and following healthy practices. Tobacco usage is associated with a number of cancers of the mouth, throat, and respiratory system. Lung cancer related to smoking is now the leading cause of death from cancer for both men and women. Since 1980, the death rate from lung cancer has increased 41 percent for both black and white women and by 11 percent for black men; the figures for white men have remained stable. Recently much attention has been given to the harmful effects of secondary exposure to smoking. Many states, as a result, have implemented legislation to limit or ban smoking in public places. See Box 9–4 for suggestions for quitting smoking.

Excessive alcohol use has been implicated in cancer of the throat, esophagus, mouth, and liver.

Many carcinogens are found in foods, especially those that are pickled, smoked, or cured. Pesticide residues left on fruits and vegetables or in meats may further place an individual at risk for developing cancers. Fat in the diet may act as a cancer promoter, causing cancer of the breast and colon. Fiber is the indigestible material contained in certain foods. It has been shown that a high-fiber diet may help prevent cancer of the colon or rectum. Fiber is found in whole grains, breads, cereals, and vegetables.

Viruses may also act as carcinogens. Some of these viruses are spread through sexual contact. Safe sexual practices, including the regular use of condoms, help prevent the spread of these viruses as well as HIV and hepatitis virus.

Other forms of cancer may be attributed to carcinogens present in today's industrial society. Many regulatory agencies have worked to help reduce the amounts of toxic materials present in the environment. Household and garden products are just two of the groups of substances that must be used just as the manufacturer specifies

BOX 9–4

Suggestions for Quitting Smoking

- Behavior modification
- Hypnosis
- Cold turkey
- Individual counseling
- Nicotine patches
- Acupuncture
- Commercial filters that gradually reduce the tar and nicotine content
- Gradual withdrawal
- Support groups

to avoid undue harm. There are safe, inexpensive, nontoxic substances that can be substituted for common toxic cleaning and insect-control products (e.g., boric acid instead of RAID; baking soda and vinegar products instead of Drano, oven cleaners, and so forth.

Over 500,000 Americans develop skin cancer during their lifetime. The main cause of skin damage and cancer is the ultraviolet rays of the sun. Everyone, especially those who are light-skinned, should avoid excessive exposure to ultraviolet light. Clothing and sun-blocking agents offer the best form of outdoor protection. A sunscreen with a **sun protection factor (SPF)** of 15 or more is generally recommended. The SPF rating is the time in minutes that a person can stay exposed to sunlight without burning.

Sensory Impairment Caused by Accidents

Young adults often take their sensory functions for granted. But lack of care and accidents can lead to sensory losses. For example, 90 percent of all eye injuries occurring in the workplace could be avoided with the use of protective eyewear. Individuals should wear protective eyewear when doing chores and repairs around the home (e.g., trimming hedges, using power tools or chemicals) or engaging in certain sports (e.g., baseball, racquetball, tennis). Eye injuries can result from chemical splashes, flying debris, or a ball traveling at high speed.

Hearing loss due to excessive noise exposure continues to be a concern for the young adult both in the home and at work. The same preventive measures are recommended for the adult as for the adolescent (see Chapter 8).

Routine eye and ear examinations can help detect cataracts, glaucoma, and hearing loss. Early detection and prompt intervention can reduce further loss of function.

Obesity

Approximately 30 percent of the adult American population is obese. **Obesity** is defined as having 20 to 30 percent excess weight. Studies have shown a direct relationship between obesity and mortality. Furthermore, obesity increases the likelihood of developing of hypertension, diabetes mellitus type 2 (noninsulin-dependent), and high cholesterol levels. High cholesterol may contribute to the onset of heart disease and strokes. Obesity has also been implicated in other conditions, such as gallbladder disease, cirrhosis of the liver, kidney disease, and some cancers. Excessive weight adds stress to the weight-bearing joints and may lead to osteoarthritis and back problems. Regular paced exercise can improve cardiovascular functions, promote weight loss, and reduce stress.

Weight loss may be accomplished through diet and exercise. Certain foods contain more calories than others: 1 g of fat yields 9 kcal; 1 g of protein or carbohydrate yields 4 kcal. Therefore people trying to lose weight may benefit from a low-fat diet. Crash diets or very low calorie diets are not only ineffective but may be harmful to one's health. Diets of this sort do little to permanently change a person's eating behaviors. Pounds lost on a crash diet are usually quickly regained. Crash diets may lead to food cravings and bingeing and set the stage for the onset of eating disorders. Crash dieting may lead to *weight cycling*—large fluctuations in weight. Recent research has shown that weight cycling often leads to a gradual increase in weight over time. Successful weight control programs are based on helping people develop lifelong behavior changes and eating habits.

Stress

Common causes of adult stress include work, marital problems, child-care concerns, and money worries. Stress reactions are highly individual and develop over years. Adults may develop certain health problems related to stress on the job or in their relationships or to their lifestyles. Sometimes as people search for career advancement and social acceptance, they may neglect health-promoting activities. Many adults pay too little attention to their nutrition and diet. Others "party" and engage in risky behaviors. Unhealthy practices during the adult years can have a direct effect on health in the later years.

Stress-management workshops can help individuals learn how to better handle or reduce stress. Most stress-reduction programs are designed to help adults learn how to manage time effectively, say no, and deal directly with the source of their stress. A sense of humor and the ability to practice relaxation are two other measures helpful in managing stress (see Fig. 9–4). Refer to Chapter 1 for other stress-reducing exercises.

Family Planning

Reproductive planning includes decisions about having children. Thanks to modern science and research, many contraceptive choices are available to the individual. Nevertheless, family planning and contraception are very controversial subjects in the United States today. The high numbers of unwanted pregnancies and elective abortions point to the need for better education and reproductive counseling services. Through education and counseling individuals are better able to make responsible decisions that are right for them and will result in happiness. Contraceptive methods should be based on the individual's values and beliefs, as well as on a given contraceptive product's reliability, side effects, and impact on sexual satisfaction.

TABLE 9–2
BIRTH CONTROL METHODS

Method	Advantages	Disadvantages
Natural, calendar, or rhythm method: monitor basal body temperature, cervical mucus for fertile and infertile times; practice coitus interruptus	Free Safe Acceptable to most religions	Must abstain for 5 days during fertile period Not very reliable
Hormonal: oral contraceptives, implant, or morning-after pill	Almost 100% effective when used properly	Weight gain Irregular menses Hypertension, increased risk for strokes, heart disease, and breast cancer
Mechanical barrier: condoms, diaphragm, cervical cap, sponge	Inexpensive May prevent transmission of STDs	May tear or dislodge Decreases sensation Increases risk of toxic shock syndrome
Chemical barriers: spermicidal creams, jellies, foam, and vaginal suppositories	Easily obtained	Messy May cause local irritation
Intrauterine devices: intrauterine progesterone contraceptive (Progestasert), intrauterine copper contraceptive (ParaGard)	Doesn't affect hormonal cycle or interrupt sex act	Infection, hemorrhage, perforation, spotting

The ideal form of birth control is one that is safe, effective, affordable, and acceptable to the parties using it. The common methods of contraception used today include hormonal methods (oral contraception, "the pill," subcutaneous implants); intrauterine devices; and chemical and barrier methods (condoms, diaphragms, spermaticides, and cervical caps). For religious or other reasons many people choose not to use any birth control devices but rely on a natural method of pregnancy prevention called the *rhythm method*. This method requires that the woman monitor her basal body temperature for fertile and infertile periods and that the parties refrain from sexual intercourse on the days of the menstrual cycle when the woman is most likely to conceive. A summary of the common birth control methods and devices is listed in Table 9–2. Sterilization via tubal ligation for women and vasectomy for men are irreversible forms of birth control that should be undertaken only by those individuals who have been counseled and fully understand the permanence of their decision.

SUMMARY

1. Early adulthood covers the period from age 20 through the early 40s. This is generally described as a stable period of growth.
2. Physical growth is completed. Most individuals have reached peak efficiency during the early portion of this stage.
3. Muscular strength and coordination peak in the 20s and 30s and then slowly decline between ages 40 and 60.

4. Gingivitis affects many adults and is considered preventable. Wisdom teeth make their appearance from late adolescence through early adulthood.
5. Early changes may be noticed in sensation and perception. After age 40, there may be a decreased ability to see objects at close distance. This condition is known as presbyopia and is treated with corrective lenses. The adult may detect some loss of high-frequency hearing.
6. The reproductive organs are functioning at peak efficiency during this stage. In women the menstrual cycle is usually well established. Women with irregular cycles or menstrual discomfort should seek medical advice. Generally men are free of reproductive problems during this stage. If problems with infertility arise, the couple should be referred to their physician for testing and guidance.
7. Cardiac changes include a gradual decline in cardiac output and a loss of elasticity in the muscles and vessels. These changes may contribute to an increase in the blood pressure. Peak respiratory function occurs during the 20s.
8. The major developmental milestones for this age group include choosing and establishing a career, fulfilling sexual needs, establishing a home and family, expanding social circles, and developing maturity.
9. Erikson viewed the psychosocial task for the adult as intimacy. He described a broader meaning of intimacy between lovers, parent and child, or friend.
10. Formal operational cognitive functioning develops further during the adult period. Adults are capable of being objective and of looking at issues from a wider perspective.
11. Adults are at the postconventional stage of moral development. As with other developmental issues, moral development progresses at a highly individualized rate.
12. Diet is crucial to health. Caloric intake is based on the adult's age, body size, amount of physical activity, and gender. Men generally need between 2700 and 3000 calories per day, while women need 1600 to 2100 calories per day.
13. On average, the adult needs 7 to 9 hours of sleep each night. Many factors may contribute to the problem of insomnia, including stress, diet, fatigue, and poor health.
14. Physical fitness can improve at any age with regular participation in exercise.
15. Accident prevention continues throughout the adult years. Improper use or care of the sensory organs can lead to disease or injury. Protection against injury during sports includes training and the use of protective clothing.
16. Concerns about safety widen to include concerns not only for oneself but for one's children and family members.
17. Yearly visits to the physician are recommended.
18. Cancer prevention is very important. Many cancers can be prevented by avoiding carcinogens in the environment and by practicing healthy living. Excessive exposure to ultraviolet rays especially in light-skinned persons may produce cancer of the skin. Sun blocks and protective clothing when outdoors may prevent skin cancer.

19. Weight management is very important. Obesity can lead to many disorders such as diabetes, heart disease, and strokes. The best approach to weight control is through education that leads to a change in lifetime diet and exercise patterns.
20. Stress management can help people learn how to handle stress more effectively by learning to handle time, saying no, and dealing directly with the stress.

CRITICAL THINKING

Vivian Andrews is your next-door neighbor. Vivian is 30 years old; she is 5 feet 4 inches tall and weighs about 180 pounds. She works part-time as a receptionist in an office. Her job is located about half a mile from her home; however, she rides the bus to and from work. She confides in you, a licensed vocational nurse, about her history of obesity. She shares with you a weight reduction diet that she has cut out of the newspaper. The diet consists of fruits and vegetables and only one protein food source each week. She has read that this will guarantee a weight loss of 10 to 15 lb in a week's time.

1. What would you advise Vivian about this diet? Give reasons to support your advice.
2. Outline a diet for Vivian that would be conducive to healthy weight loss.
3. List several lifestyle changes or modifications that would be health-promoting for Vivian.

Multiple-Choice Questions

1. Cynthia Beckford is a healthy 30-year-old woman. Which of the following normal age-related physical changes would you expect her to be experiencing?
   a. Increase in bone cells
   b. Decrease in muscle cells
   c. Increase in new brain cells
   d. Loss of some elasticity in the lung

2. The major cause of tooth loss in individuals over the age of 35 years is:
   a. Tooth density
   b. Dental caries
   c. Gingivitis
   d. Stomatitis

3. Andrew Previs, age 40, is having his annual eye examination. He has noted a decline in his visual acuity. The most likely cause of this symptom at this age is:
   a. Widening of the iris
   b. Eyestrain
   c. Opacity of the lens
   d. Loss of corneal cells

4. The psychosocial task for the young adult is:
   a. Identity
   b. Intimacy
   c. Introspection
   d. Egocentricism

5. The psychological outcome of prolonged unemployment is often:
   a. Job phobia
   b. Social isolation
   c. Regressive behavior
   d. Selflessness

6. A sign of maturity in adulthood is the individual's ability to:
   a. Exert excessive self-restraint
   b. Develop an interest in community activities
   c. Make life choices based on the advice of others
   d. Frequently change jobs

Suggested Readings

Bakken, L, and Ellsworth, R: Moral development in adulthood: Its relationship to age, sex, and education. Ed Res Q 14(2):2–9, 1990.

Chassin, L, Presson-Clark, C, Sherman, S, and Edwards, D: The natural history of cigarette smoking and young adult social roles. J Health & Social Behavior 33(4):328–347, 1992.

Freysinger, VJ: The dialectics of leisure and development for women and men in mid-life: An interpretive study. J Leis Res 27(1):61–84, 1995.

Robinson, L, Garthoeffner, J, and Henry, C: Family structure and interpersonal relationship quality in young adults. J Divorce & Remarriage 23(3/4):23–29, 1995.

Santrock, J: Life-Span Development. Brown & Benchmark, Madison, WI, 1995.

Schuster, C, and Ashburn, SS: The Process of Human Development: A Holistic Approach. JB Lippincott, Philadelphia, 1992.

U.S. Department of Health and Human Services: Healthy People 2000: National Promotion and Disease Prevention Objectives. DHHS Publication No. (PHS) 91-50212, 1992.

Middle Adulthood

Chapter 10

Chapter Outline

Middle Adulthood

Key Words

benign prostatic hypertrophy
cataracts
climacteric
coitus
dermis
dyspareunia
empty-nest syndrome
fibrocystic breast disease
generativity
glaucoma
hormone replacement therapy (HRT)
hot flashes
menopause
presbycusis
procreation
stagnation

Learning Objectives

At the end of this chapter, you should be able to:

- List three physiological changes that occur during middle age.
- Describe the psychosocial task that Erikson identified for this stage.
- List three goals unique to this stage of development.
- Describe three areas of health concern for the middle-aged adult.

Middle adulthood, or middle age, covers the period from the mid-40s through the early 60s. In the past, middle adulthood was defined as the period of development after traditional childbearing roles were completed. Today, however, many women are entering the workforce, delaying marriage and childbearing in order to advance their careers. For this reason the more current definition of middle adulthood is a transitional period of development after the early adult years but prior to the retirement years.

There are many conflicting images about the meaning of middle age. Some describe this stage as the peak of life; middle-aged individuals are often considered to be powerful, wise, and in control. Conversely, there is the belief that life is downhill after 40; middle age is sometimes portrayed as a time when there is a decrease in energy and physical attractiveness, along with an unhappy home life.

Our belief is that middle age is a natural consequence of development and should be viewed as a time of growth and progression rather than of decline and regression. During this stage of development adults should reach their peak performance and maturity if they maintain a healthy lifestyle.

Several goals have been identified for this stage, including establishing and adjusting to new family roles, securing economic stability for the present and future, maintaining a positive self-image, and evaluating or redesigning career options (Fig. 10–1). In addition, adults must maintain a healthy lifestyle and physical well-being.

PHYSICAL CHARACTERISTICS

Height and Weight

As individuals age they may gradually lose 1 to 4 inches in height. Body contour changes because of an increase in fat deposits in the trunk region. Middle-aged adults also note that, even without gaining weight, they may require a larger clothing size because their body weight becomes redistributed. Proper exercise and diet can help slow down the effects of aging on these body systems.

FIGURE 10–1
The middle-aged couple must adjust to new family roles.

Muscle and Bone Development

Most physiological changes associated with this stage appear gradually and at different times for different persons. A noticeable change during this stage is related to a loss of muscle tone and elasticity in the connective tissues. This gives the skin a flabby, less firm appearance in the muscles of the face, abdomen, and buttocks. As muscles decrease in tone, there is a gradual decrease in muscle strength. Bones lose mass as a result of demineralization. This results in more porous, brittle bones.

Dentition

Some adults complete their dentition, while others may not have room for their third molars (wisdom teeth) to fully erupt. If these impacted wisdom teeth are causing pain and problems, some individuals may require surgery. Periodontal disease can be prevented with proper mouth care and maintenance (see Chapter 9). Regular check-ups, brushing properly and flossing have been shown to help maintain healthy teeth and gums.

Development of Other Body Systems

The muscles of the heart and lungs lose elasticity and there is a slight decrease in maximum efficiency: an increase in blood pressure and a decrease in the efficiency of air exchange. In general all muscles show a slight decreased capacity to perform work and require a longer time to recover after exertion.

The most obvious changes are in the skin. The cells of the **dermis** (inner layer) become less elastic, resulting in wrinkling and sagging of the skin. The most obvious changes appear on the face, causing laugh lines, lines around the eyes, and a loosening of the skin under the chin. Marked weight loss can further exaggerate these changes. Noticeable hair changes for some individuals include graying, thinning, and slowed growth. Race, genes, and gender all influence patterns of hair growth.

There is a decline in both visual and aural acuity (see Chapter 9). The lens becomes thicker and more opaque, leading to a decrease in peripheral vision. The eyes have a decreased ability to focus on near objects (presbyopia) and are less able to adjust as well to changes from dark to light. Most of the visual changes can be successfully managed with the use of corrective lenses. New laser surgery can correct some of these defects with a rapid recovery period. Many people develop a degree of **presbycusis**, or loss of hearing acuity. This is related to a thickening in the walls of the capillaries in the ear. Lack of proper care and excessive exposure to loud noise can further exaggerate these losses.

During this stage adults may begin to notice a gradual loss of taste discrimination. This usually is a minor change which doesn't affect appetite or food selection. The general condition of the mouth and teeth will more often affect diet.

VITAL SIGNS

There should be no significant changes in the healthy middle-aged adult's vital signs as compared to the early adult period. Refer to Chapter 9 for normal adult ranges.

DEVELOPMENTAL MILESTONES

Sexual Development

People continue to live as sexual beings throughout their entire lives, although advancing age may change the person's options, opportunities, and means of sexual expression. Many myths still exist with regard to middle-aged sexuality and performance. The middle years are often depicted as a period of decreased sexual activity, pleasure, and interest. Some even portray middle-aged marriage as a loveless, sexless relationship. For many people today middle age is a time when both partners are actively engaged in the workforce, seeking financial and career rewards. This may leave little time to nurture intimacy in sexual relationships. Many couples complain that they have little time or energy for their sexual relationships.

The loss of reproductive capacity, either naturally or due to surgery, is no cause for the loss of libido (sexual drive) and sexual pleasure. Very often women are at their peak sexual capacity, desire, and pleasure during this period. Sexual concerns for men relate to changing roles, lowered levels of testosterone, and anxiety over sexual performance. Other psychosocial factors, including work-related stress and a general lack of physical fitness or health, may cause men to note a diminished sexual responsiveness.

The change of life that occurs in both men and women is scientifically referred to as the **climacteric**. In women this change of life is most commonly known as **menopause**, the cessation of menses, which results from a progressive decrease in the production of estrogen and progesterone. The completion of menopause results in the end of reproductive ability. There appears to be a strong genetic influence affecting the onset and duration of menopause. That is, daughters can generally predict the course of their change of life by looking at their mother's menstrual history. Menopause usually begins between 45 and 55 years of age, with the average age of onset being 51.

Menopause usually begins with noticeable changes in the woman's menstrual cycle. The cycle may become irregular and shorter in duration, with longer intervals between periods. Some women experience spotting between periods. The amount of blood flow may increase or decrease (normal blood loss during menses is between 30 and 60 mL), or the menses may come to a sudden stop. Internally, the ovaries, fallopian tubes, and uterus decrease in size. The ovaries no longer secrete reproductive hormones and ova. The vagina may lose some elasticity and become drier. These changes may lead to itching and discomfort (**dyspareunia**) during **coitus** (intercourse). Using a lubricant during sexual intercourse will lessen the discomfort caused by increased vaginal dryness.

Menopausal women frequently experience **hot flashes**, which are caused by vasodilatation of the capillaries and a sudden rush of blood to the skin surface. During the hot flash the body becomes very warm; this is followed by excessive perspiration, vasoconstriction, and chilling. The hot flash involves mostly the head and neck region and may be very visible to the onlooker. It may last for a few seconds or up to a few minutes and may reoccur any number of times a day. Frequently the individual may complain of other symptoms as well, including night sweats, insomnia, and a general feeling of anxiety. Table 10–1 summarizes menopausal signs and suggested interventions.

Hormone replacement therapy (HRT) is an effective treatment for lessening menopausal symptoms. HRT is not recommended for women with a family history of

TABLE 10–1
MENOPAUSAL SIGNS AND SUGGESTED INTERVENTIONS

Signs	Interventions
Hot flashes followed by chills	Dress in layers Avoid high necklines Wear cotton
Palpitations, nervousness, headache	Routine physical examination to rule out medical problems Decrease stress
Loss of muscle strength	Regular exercise
Decreased elasticity of the skin	Limit exposure to the sun
Increased facial hair	Remove with electrolysis or waxing
Decreased vaginal lubrication	Use water-soluble lubricants before coitus
Sleep disturbances, fatigue	Follow relaxing routines at bedtime

breast disease or cancer. Hormone replacement therapy has also been reported to lessen the risk of heart disease and osteoporosis when used under proper medical supervision.

Some women view menopause as a natural event that will culminate in new freedom and beginnings. Others may view menopause more negatively—as an end to their reproductive years. Education and an exploration of their feelings can help women gain insight into and understanding of their bodies and the changes that they experience at this stage.

There is no significant physiological change of life for males. During middle age there may or may not be lower levels of testosterone with fewer numbers of viable sperm. Men remain capable of producing sperm and of **procreation** (reproduction) well into their 80s. The main change that men experience during middle age relates to their thinking patterns and self-image. As they notice some of the physical changes associated with middle age, some men try to look and act younger. Problems with male sexual functioning are usually caused by disease or related to mental outlook. Middle-aged men who are threatened by the aging process may find that they are also having difficulties functioning sexually. Some may even attempt to prove their sexual appeal by taking on younger sexual partners.

PSYCHOSOCIAL DEVELOPMENT

According to Erikson, the primary the task of middle adulthood is **generativity**, which refers to an individual's desire and ability to serve the larger community. For many adults, generativity also includes having a positive influence on their own children. However, as Erikson points out, an adult does not have to have children to be generative. With or without children, many middle-aged adults are more self-confident about the skills and knowledge they have acquired over the years. Thus, they may feel more able to expand their nurturing beyond their immediate family circle. They often demonstrate a concern for their community and look to what they can do to make improvements that will benefit future generations. Examples of generativity include volunteer work in the church, school, hospital, or community. Achievement of their own lifelong goals and of larger generative goals usually results in satisfaction about themselves and their lives.

Failure to achieve generativity results in self-absorption and **stagnation**. These individuals are preoccupied with themselves and refuse to accept life as it is or make changes in things with which they are dissatisfied. Immaturity and self-absorption may lead to depression and acting-out behavior.

Establishing and Adjusting to New Family Roles

Role changes are a part of middle adulthood, as they are in other life stages. For many, time is no longer primarily spent on child-centered activities, but on couple-centered activities. The couple may find more leisure time to devote to pursuing their own interests. Those couples who have completed their early-parenting roles may find that they now have to reacquaint themselves with each other and redefine their new roles and responsibilities.

To successfully make the transition required by new roles, each partner needs to appreciate the other's growth, individuality, and needs. Both individuals in a relationship must be flexible and ready to support each other's struggle to adapt to their new roles. Although respect for the partner's individuality is needed at any stage in a couple's relationship, marriages in middle age often show signs of stress caused by earlier unresolved conflicts. These conflicts may be related to finances, role division, and intimacy. Problems that resurface, as well as new problems that may arise, must be discussed and resolved in order to preserve the relationship. Some couples may seek marriage counseling if they need additional outside help in resolving conflicts. Research shows that fewer divorces take place among those in middle age than among couples in their 20s and 30s. Most commonly, divorce occurs during the first 3 years of marriage.

One adjustment that may be particularly stressful for some couples occurs when the last child has left home. The **empty-nest syndrome** is sometimes especially difficult for women. That is, individuals whose identity revolves around few activities other than parenting may have a more difficult time adjusting to a home without children. This phenomenon is less common today than it was in an earlier generation because more women manage dual roles—in the home and in the workplace. Many middle-aged women find that there is now less stress and more time to pursue their own goals and ambitions.

Others may have postponed marriage and childbearing and now, in their middle years, still have young children at home. One result of this trend is a decrease in the average number of children per family. That is, the longer marriage is postponed, the fewer children the couple is likely to have after marrying. Parenting at this stage may place additional stress on the parents because their energy levels may not be quite as high as when they were younger. On the other hand, young children may bring a new youthfulness and spurt of energy to the lives of middle-aged parents.

Many people become grandparents in midlife. Today's grandparent is very different from those of the past. No longer is this role solely associated with advanced age and infirmity. Many of today's grandparents are youthful in their appearance and outlook. Some are still actively working or fulfilling lifelong dreams. Others assume the child-care role, allowing their grown children, the parents, to work or complete their education. Still others find pleasure and joy in the role of grandparenting and leave the act of primary care to the child's parents.

Just as middle-aged individuals' relationships with their children change, so do their relationships with their parents. Most middle-aged adults have close relationships with their parents and maintain regular, frequent contact. Some continue to have a mutually nurturing relationship, while others maintain a relationship based on a sense of duty and obligation.

Middle-aged adults may find that they need to adjust to the role of parenting their parent. During middle age some adults realize that they can no longer rely on their parents for support; instead their parents now need them for support. Economic problems or failing health may result in the need for a change in roles. Caring for elderly parents can cause added stress on the family. If elderly parents need care, more often the daughter assumes this role. For many middle-aged adults, caring for elderly parents is a major challenge. Sometimes decisions need to be made about helping parents to move to retirement centers or nursing homes, or to make their home with their middle-aged children. All feelings accompanying these changes must be acknowledged. Individuals caring for elderly parents need to balance or arrange for outside support so that they, too, get respite from caregiving. See Box 10–1 for hints on caring for aging parents.

BOX 10–1

Caring for Aging Parents

- Recognize and respect older parents' feelings.
- Expect ambivalent feelings—both a sense of responsibility and a sense of resentment.
- Maintain open communications between siblings and other family members.
- Establish limits and delegate tasks whenever possible. Include pleasurable activities in your daily activities.
- Seek out support services, support groups, home health care, respite care, and senior-citizen centers.

Establishing Economic Security for the Present and Future

By middle age, most people are at their peak earning capabilities and job status. Plans for economic security best begin when people first enter the job market. It is not sufficient to wait until middle age to start to save toward retirement.

Economic security may be strained when middle-aged parents are paying or helping to pay for their children's college education. College has a major impact on many middle-aged parents' financial security. If parents have delayed having children, the need to finance their children's higher education may come when their own earning capacity is beyond its peak. Economic security also becomes a special challenge for middle-aged adults who need to help their own parents financially.

Maintaining a Positive Self-Image

The development of true intimacy is critical to the survival of close relationships. Intimacy promotes trust and mutual caring. The deepening of intimate bonds can allow partners to share their joys and defeats in a mutually supportive, enabling manner.

Middle-aged adults need to accept the visible age-related changes of this stage without a loss of self-esteem. In our youth-oriented society so much emphasis is placed on the importance of looks and staying young that individuals may feel threatened by the aging process. Many resort to surgery, cosmetics, dieting, and exercise to preserve their youthful appearance. A balance of mental and physical health and positive social interactions will allow the individual to maintain a healthy sexuality as well as general positive self-esteem.

Evaluating and Redesigning Career Options

Adults hope to reach their peak career goals by middle age. Those persons who have not done so must come to terms with their accomplishments. This may result in the decision to change careers or go back to school. The concept of a single job or career may be a thing of the past. Most adults today work more than one job in order to meet the rising costs of living. Some women enter the workforce for the first time when their children are grown and more independent. Others find that they are forced to make job changes because of changes in technology and the job market. Job loss, retraining, and relocation all have an impact on today's middle-aged workers.

Cognitive Development

Mental ability and memory remain at peak performance, as in the earlier adult period. Middle-aged adults are capable of thinking in a pragmatic and concrete manner. They display a unique potential to integrate objective and rational modes of thinking; these are signs of true maturity. Many middle-aged adults are enrolled in courses to help further their job-related knowledge or fulfill personal areas of interest. Individuals returning to school at this time may encounter some difficulty in adjusting to the learning environment. For example, they may have difficulty setting up study schedules, memorizing may be difficult, and just being in a classroom situation may create some stress for the adult. Once enrolled adult learners go through a brief period of adjustment. Following this, providing they do not get discouraged and quit, they quickly acclimate to the demands of learning. They may need more time to learn and complete tasks but often do so with much more accuracy than the young learner. Motivation to learn is often greater and is enhanced by life experiences and needs. Box 10–2 has some suggestions for the adult learner.

Moral Development

Middle age is a time in which many individuals look into themselves and reassess their values and beliefs. Spirituality may become more important during this stage

BOX 10–2

Suggestions for the Adult Learner

- If several years have lapsed since you were last in school, it may take a few weeks to get back into the routine.
- Motivation and perserverance are important keys to success in learning.
- Keep up with the reading assignments; ask yourself if you understand what you have read.
- Do the learning objectives in the text.
- Look up words that you don't understand.
- Study the illustrations and tables given in the text.
- Summarize what you read.
- Complete the end-of-chapter questions and exercises.
- Prepare for your examination from the first day of class.
- Go over your test after it is scored to figure out what you missed. Use these errors as clues to what you must review for future exams.

and may guide the person in making moral decisions. A commitment to improving the welfare of others enhances the individual's own moral growth. At this stage of development, most people have a clear understanding of what constitutes personal needs, moral duties, and society's demands.

NUTRITION

The dietary needs of the middle-aged adult are similar to the needs of those in the early adult stage. The basal metabolic rate gradually slows down during the middle years, possibly affecting the weight of the individual. It is not unusual to gain 5 to 10 lb and go up a clothing size without any increase in food intake. This is related not only to the slowing of the metabolism but also to a redistribution of the individual's weight. To compensate for these changes, middle-aged adults must decrease their caloric intake and increase their amount of physical activity. Healthy eating patterns should continue throughout this stage. Adequate calcium is needed and can be achieved by ingesting two or more servings of calcium-rich foods per day. (See Chap. 9 for specific dietary guidelines.)

SLEEP AND REST

Sleep requirements for middle-aged adults are less than for people in earlier stages. Many individuals complain that they experience more difficulty falling asleep or staying asleep. Stress, poor health, or lack of exercise may contribute to sleep problems. Many middle-aged adults notice that they no longer have an abundance of energy. They may tire more easily and need rest periods following strenuous exercises. (See Chap. 9 for ways to promote sleep.)

EXERCISE AND LEISURE

Some middle-aged adults are sedentary and need to be reminded of the benefits of exercise, whereas others actively engage in regular exercise (Fig. 10–2). Chapter 9 outlines different types of exercise. Choice of hobbies varies greatly among individuals. For many adults, leisure activities may center around the home. Travel, gardening, art, and music are just a few of the areas of interest for some in this age group (Fig. 10–3). Some adults develop new interests or talents, whereas others now have the time to nurture old interests. Many persons find pleasure in devoting time and service to others in their community. Volunteering at local hospitals and schools benefits both the doer and the recipients. Filling leisure time with rewarding, pleasurable activities is important in preparing for retirement. When the leisure activities of midlife are continued into the later years, there is a smoother transition into old age.

SAFETY

Accident prevention is a concern at all ages. Motor vehicle accidents continue to contribute to many injuries and deaths. Identifying safety issues and reducing risk factors in the workplace helps to decrease the number of job-related injuries and accidents. The Occupational Safety and Health Act of 1970 was passed to increase the

FIGURE 10–2.
Exercise and leisure activities help maintain a healthy outlook.

FIGURE 10–3.
Travel and leisure are important at this stage of development.

health and safety of all working men and women. Continuous surveillance and identification of individuals at high risk for injury helps promote preventative strategies.

HEALTH PROMOTION

It is completely possible for a person to maintain sound health throughout the middle years despite the general slowing down of body processes. Weight control, healthy lifestyles, and the avoidance of accidents helps to promote wellness into the later years. Many of today's middle-aged adults are concerned with physical fitness. Much media attention has been devoted to diet and health care, making many members of this age group very conscious of wellness and healthy lifestyles.

In order to maintain health and wellness, the middle-aged adult must have a yearly physical examination. Middle-aged adults should also have eye screening tests for glaucoma and cataracts. **Glaucoma** is a condition that frequently begins in middle adulthood. It is caused by the buildup of fluid in the chambers of the eye. This increased pressure may go unnoticed until irreparable damage has been done to the person's vision. A routine noninvasive eye exam may help detect the onset of glaucoma. Medications or corrective surgery can help prevent further loss of vision. **Cataracts** result in a cloudy formation on the lens of the eye. Cataracts form gradually and eventually inhibit the passage of light through the lens. They are common after the age of 60 and may be caused by the presence of other chronic conditions. Once detected, cataracts can be surgically corrected with excellent results.

Cancer of the colon is seen more frequently during middle adulthood than at earlier ages. It is therefore suggested that both men and women have periodic proctoscopic examinations after the age of 50. Much attention has been given to diet and its relationship to colon cancer. A diet high in fiber and low in fat is recommended to decrease the risk of both colon cancer and heart disease.

Fibrocystic breast disease, a benign condition, may begin at age 30 and continue until menopause. Ninety percent of all women develop some degree of this disease, which is characterized by large, sometimes uncomfortable cysts in the breast tissues. Estrogen has been implicated in contributing to the development of this disease. It is thought that estrogen helps to cause engorgement and swelling of the cells lining the mammary ducts. After menopause, with the decrease in estrogen levels, there appears to be a decrease in tissue swelling and cyst formation. At this time there is no known correlation between the development of fibrocystic disease and the onset of breast cancer. However, all irregularities and lumps found in breast tissue should be investigated thoroughly by a trained examiner. Women over 50 should have annual mammograms and Pap screenings.

Men over 50 frequently develop an enlarged prostate gland. This condition, called **benign prostatic hypertrophy**, is common and should be distinguished from cancer of the prostate gland. Early signs of this condition include difficulty in voiding, diminished urinary stream, dribbling, and frequency. A yearly rectal exam after age 40 will help detect an increase in the size of the gland. Blood testing can help detect prostatic-specific antigens (PSAs) found at an early stage of prostate malignancies. Early detection and prompt treatment can improve the outcome of both conditions.

Heart disease and cancer continue to be the cause of most deaths at this stage of development. The leading chronic conditions affecting middle-aged men include diseases of the heart, back problems, visual impairments, and asthma. The leading chronic conditions affecting women include arthritis, hypertension without heart involvement, and depression.

SUMMARY

1. Middle adulthood or middle age covers the period from the mid-40s through the early 60s.
2. Today many women entering the workforce delay marriage and childbearing to enhance their careers.
3. Middle age is a natural consequence of development and a time of growth and progression rather than decline and regression.
4. During this stage adults reach their peak performance and maturity.
5. Most physiological changes appear gradually and at different times for different persons.
6. There is a loss of muscle tone and elasticity in the connective tissues. There is also weight gain. Demineralization causes the bones to become porous and brittle.
7. Periodontal disease is common and can be prevented with proper mouth care and maintenance.
8. The skin loses elasticity and becomes wrinkled.
9. Hair growth slows; thinning and graying of the hair occur.
10. The eyes lose accommodation and the ability to focus on near objects. A loss of hearing acuity is noticeable at this stage.
11. Goals at this stage include establishing and adjusting to new family roles, securing economic stability for the present and future, maintaining a positive self-image, and evaluating or redesigning career options.
12. The major role changes that middle-aged adults experience now include couple-centered focus, empty-nest syndrome, grandparenting

or new parenting, and parenting the parent. Adults hope to establish economic security and reach peak job status during middle age.

13. Menopause indicates the cessation of menses and a loss of reproductive ability. Menopause usually begins between the ages of 45 and 55 years.
14. There is no significant physiological change of life for men. Men remain capable of producing sperm well into their 80s. The main change that men experience relates to their thinking patterns and self-image.
15. According to Erikson, the task of middle age is generativity. This means that individuals demonstrate a concern and interest in their community. Nonachievement of generativity results in self-absorption and stagnation.
16. Mental ability and memory remain at peak level of performance. This is a time when adults look into themselves and reassess their values and beliefs.
17. Middle-aged adults have a clear understanding of what constitutes personal needs, moral duties, and society's demands.
18. The nutritional needs of the middle-aged adult remain similar to the needs of the young adult. This age group must pay close attention to diet, exercise, and healthy lifestyles. Middle-aged adults require less sleep than at earlier stages. Some adults experience difficulty falling or staying asleep.
19. Leisure-time activities vary and are important in preparing for retirement.
20. Heart disease and cancer continue to be the leading causes of death for this age group. Thorough physical examination and health screening must be performed yearly to help detect and treat any existing medical problems.

CRITICAL THINKING

Mary Jo Frazer, a 52-year-old postal clerk, is having her yearly physical examination. She tells the nurse that during her last menstrual period she noted that she had 2 days of increased blood flow and 2 to 3 days of increased spotting before her period ended. She also complained about having periods of increased anxiety and night sweats.

1. How should the nurse respond to her concern?
2. What screening test is mandatory for this patient?
3. What patient teaching is indicated at this time?

Multiple-Choice Questions

1. Wrinkling of the skin seen during late middle age is due to:
   a. Increased water and decreased fat in the skin cells
   b. Loss of elasticity in the dermis

c. Increased muscle mass and stretching of fibrous tissue
d. Rapid loss of cells from within the dermis and epidermis

2. The following are characteristic of middle age:
   a. Women are capable of giving birth well into their 60s.
   b. Men are incapable of producing sperm after the age of 70.
   c. Sexual needs and desires cease.
   d. Sexual functioning and sexuality continuing throughout this stage.

3. Hot flashes are caused by:
   a. Nervous system excitement
   b. Hormonal influx
   c. Vasodilatation and constriction
   d. Decreased contractility in the blood vessels

4. The psychosocial task of generativity refers to:
   a. How one chooses to achieve economic stability
   b. The task of procreation
   c. Accomplishing one's career and ambitions
   d. Assisting and guiding the next generation

5. Successful coping with the midlife changes are best when the individual:
   a. Has children of his or her own
   b. Is married
   c. Has a career
   d. Has a good support system

6. Cynthia Fox is a 55-year-old woman who has recently been diagnosed with fibrocystic breast disease. She asks the licensed vocational nurse (LVN) if this disease occurs because she didn't breast-feed her two children. The LVN's best response is:
   a. "Don't worry, nothing you could have done would result in this disease."
   b. "There is nothing that you can do at this time to halt the course of this condition."
   c. "This disease results from hormonal stimulation of the breast tissue."
   d. "Breast-feeding your children usually will decrease the risk of this disease."

Suggested Readings

Bakken, L, and Ellsworth, R: Moral development in adulthood: Its relationship to age, sex, and education. Ed Res Q 14(2):2–9, 1990.

Finkel, D, Pedersen, N, and McGue, M: Genetic influences on memory performance in adulthood. Psychol Aging 10(3):437–446, 1992.

Hartweg, DL: Self-care actions of healthy middle-age women to promote well-being. Nurs Res 42(4):221–227, 1993.

Hess, BB, and Markson, EW: Growing Old in America. Transaction, New Brunswick, NJ, 1991.

Hunter, S, and Sundel, M: Midlife Myths: Issues, Findings, and Practical Implications. Sage Publications Inc., Newbury Park, CA, 1989.
Schuster, C, and Ashburn, SS: The Process of Human Development: A Holistic Approach. JB Lippincott, Philadelphia, 1992.
Wilbur, J, Dan, A, Hendricks, C, and Holm, K: The relationship among menopausal status, menopausal symptoms, and physical activity in midlife women. Family and Community Health 13(3):67–78, 1990.

Chapter 11

Chapter Outline

Late Adulthood

Key Words

ageism
aging
antioxidants
atrophy
cerumen
demographics
dysphagia
ego integrity
gerontology
homeostasis
integumentary system
keratosis
kyphosis
lacrimal ducts
life expectancy
life span
lipofuscin
lumen
melanocytes
nephrons
neurons
old
opacity
peristalsis
pruritus
reminiscence
residual volume
senescence
senile lentigo
tinnitus
very old
xerostomia
young old

Learning Objectives

At the end of this chapter, you should be able to:

- Describe three demographic changes affecting the older population.
- Contrast the biological and psychosocial theories of aging.
- List four normal physical age-related changes that occur during this stage of development.
- Describe two developmental milestones associated with aging.
- Describe Erikson's pyschosocial task for this period of development.
- List three dietary changes important for old age.
- List two health-promoting activities important for old age.

"Old age," as defined by the U.S. government and Social Security Administration, includes all those age 65 and up. The statistical characteristics, or **demographics**, of the older population are constantly changing. Old age is best divided into three periods: the **young old**, ages 65 to 74; the **old**, ages 75 to 90; and the **very old**, from age 90 on. As with earlier life stages, not everyone over age 65 is the same. Some 80-year-olds lead active, productive lives, whereas others are unable to be active or independent because of illness. Chronological age is usually an unreliable indicator of mental, physical, and social well-being.

Elderly people constitute the fastest-growing group in the United States today. Since 1990 the elderly population has increased by 7 percent, as compared to a 4 percent increase in the population under 65. In 1994 there were 33.2 million people aged 65 and over—12.7 percent of the total population. By the year 2030 there will be more than twice as many: approximately 70 million people 65 and older, or about 22 percent of the population. The majority (68 percent) lived in a family-type setting in 1994. Four out of five older persons have children who live within 30 minutes of their residence. Most have weekly visits with their children and or talk with them on the phone regularly. Only 5 percent of those aged 65 to 85 live in institutionalized settings. This number increases dramatically, to 25 percent, for those over age 85.

The elderly population is concentrated in nine states: California has 3 million; New York and Florida each have over 2 million; Pennsylvania, Texas, Ohio, Illinois, Michigan, and New Jersey each have over 1 million. Older people are less likely to change their residence than younger adults. Those who do move within a short distance from their present home.

In 1994 the major source of income for older couples and individuals was Social Security. The median income was $15,250 for men and $8,950 for women. In 1992 sources of income were as follows:

- Social Security, 40 percent
- Assets, 21 percent
- Pensions, 19 percent
- Earnings, 17 percent
- Other sources, 3 percent

In 1994 3.7 million of elderly persons (about 11.7 percent) were living below the poverty level. One fifth of the total older population were poor or "near poor." Of these, 10 percent were white, 27 percent were African American, and 23 percent were Hispanic.

Life expectancy refers to the average number of years that a person is most likely to live. The single most effective predictor of life expectancy is one's biological parents. The life expectancy for women is currently greater than for men (18.9 years for men or 83 years of age and 25.3 years for women or 90 years of age). Life expectancy has increased greatly since 1900 because of improvements in medical care for infants and young adults, in sanitation, and in overall health practices. Dietary practices over a person's lifetime may affect life expectancy. Obesity of 20 percent or more and sedentary lifestyle practices increase the risk of early death.

THEORIES OF AGING

Life span is best defined as the maximum number of years that a species is capable of surviving. Life span for humans is 120 years and has remained essentially unchanged for the past 100,000 years.

The aging process begins at conception. This process leads to physiological impairment and eventual death. **Aging** is a normal, inevitable, progressive process that produces irreversible changes over an extended period of time. It is important to note that although all persons age, they do so at a very individualized rate. The symptoms of normal aging are referred to as **senescence**. Many myths and misconceptions still exist with regard to the aging process. See Box 11–1 for a list of the common myths about aging.

BOX 11–1

Common Myths about Aging

Most old people:

- Are senile
- Live alone, isolated from their families
- Are ill
- Are victims of crime
- Live in institutional settings
- Are set in their ways and can't learn new skills
- Are unhappy
- Are less productive than younger workers
- Have no interest in sex
- Live at or below the poverty level

The study of aging is called **gerontology**. No one concept completely explains the aging process or why we age. Many different theories have been developed to attempt to explain the mysteries of aging. Most provide guidelines for assessing a person's adjustment to aging. Understanding aging helps nurses assess, implement, and evaluate care for elderly people.

Biological Theories

Biological theories attempt to explain the physical changes that accompany aging.

Clockwork Theory

Laboratory studies have revealed marked differences in cell reproduction in different species. Cells in species known to have longer life spans reproduce more times than cells of species having shorter life spans. According to the clockwork theory, connective-tissue cells have an internal clock that is genetically programmed to stop cell reproduction after so many reproductions. This "clock" determines the length of one's life.

Free-Radical Theory

Free radicals are highly unstable molecules that may result from cellular metabolism or substances found in the atmosphere. These particles are very reactive and may combine with proteins, lipids, or cell organelles. Free radicals are believed to cause mutations in the chromosomes, thereby changing cellular functions. **Antioxidants**, such as vitamins C and E, are thought to prevent the formation of free radicals and are therefore considered very important dietary substances. The exact role that antioxidants play in the aging process still remains unclear.

Wear-and-Tear Theory

This theory suggests that after repeated injury, cells wear out and cease to function. According to this theory, metabolic waste products accumulate over time. These waste products deprive the cells of their nutrition and cause the cells to malfunction.

Immune-System-Failure Theory

The immune system provides the body with antibodies and defenses against foreign invaders. The immune response declines with advancing age. The older body loses lymphoid tissue from specific locations in the body, including from the thymus gland, spleen, lymph nodes, and bone marrow. The decline in the immune functions causes the body to slow its response to foreign invaders, making elderly people more susceptible to both major and minor infections.

Autoimmune Theory

The immune system is programmed to recognize and differentiate its own proteins from foreign invaders. As the individual ages, the immune system appears to lose this ability. As a result, the body begins to attack and destroy its own cells. During old age there is an increase in the body's autoimmune response. This is evidenced by a greater incidence of autoimmune diseases such as rheumatoid arthritis and possibly cancer.

Psychosocial Theories

These theories attempt to explain how aging affects socialization and life satisfaction.

Disengagement Theory

The disengagement theory suggests that society and the individual gradually withdraw or disengage from each other. Proponents of this theory believe that disengagement provides a means for an orderly transfer of power from the old to the young, and that this process is mutually satisfying for both groups. Elderly people are relieved of their societal responsibilities and pressures, while younger people assume leadership. Critics of this theory, however, believe that as older people's level of engagement decreases, their level of contentment also decreases.

Activity Theory

This theory suggests that individuals achieve satisfaction from life by maintaining a high level of social activity and involvement. Supporters of the activity theory advise older individuals to find rewarding, pleasurable substitutes for earlier activities. They recommend that older adults remain active in a wide variety of pursuits. If activities must be given up because of age-related changes, replacements must be found. Failure to replace old activities or roles causes people to feel that they have no purpose or social importance. The activities that are most rewarding are those that involve close personal contact. By remaining active, individuals achieve a higher morale and personal adjustment than those who are less active and involved (Fig. 11–1).

Continuity-Developmental Theory

This theory views each person as a unique individual with a distinct personality. The person's personality and pattern of coping remain unchanged with aging. The aging process is seen as a part of the life cycle, not as a separate terminal stage. Personality patterns are developed over a long period of time and help determine whether the person remains active or inactive and engaged or disengaged from society. Knowledge of personality type may be helpful in predicting the person's response to aging. The individual's state of health will also determine how long he or she will remain active and satisfied. Illness may lead to retirement, social isolation, and a decrease in self-esteem.

PHYSICAL CHARACTERISTICS

Quality of life is not age-dependent but is largely determined by the person's ability to independently perform activities required for daily living, such as dressing,

FIGURE 11–1
Remaining active and involved helps to achieve a sense of personal fulfillment.

bathing, toileting, and eating. Health problems should not be viewed as inevitable, as many can be prevented or controlled.

Height and Weight

Many signs of aging are evident in both the conformation and composition of the body. Trunk length decreases as spinal curvature increases and the intervertebral disks compact. This process actually begins much earlier: on the average, adults lose 1 cm per decade after age 30. There is also a decrease in shoulder width in both sexes due to the loss of muscle mass in the deltoids. There is a slight increase in chest circumference resulting from the loss of elasticity in the lungs and in the thorax. The circumference of the head decreases and the nose and ears lengthen. Body weight decreases slowly after age 55. Other changes include a loss of body surface area and of active cell mass. Older adults have 30 percent fewer cells than younger adults. Body fat **atrophies** (shrinks), giving a bony appearance and a deepening of body areas in the axillae, rib cage, and orbital cavity surrounding the eyes. These changes in body composition are vitally important in helping to understand drug metabolism and nursing interventions for this age group. Decreased body surface area and body fat affect the dosage and rate of drug absorption. To accommodate these physiological changes, lower dosages of medication are used for older persons.

Musculoskeletal System

Postmenopausal women lose bone mass at a faster rate than men, putting them at greater risk for osteoporosis. The typical person at risk for osteoporosis is the aging, thin, white, menopausal woman. Women over age 80 have a 1 in 5 chance of sustaining a fracture of the femur. Osteoporosis is now being investigated as a health risk for men as well as for women. Regular active or passive exercise can minimize discomfort and loss of bone mass. Postural changes also occur, resulting in **kyphosis**, an exaggerated curvature of the spine, or the typical dowager's hump (Fig. 11–2). The tilting of the head and flexion of the hips and knees causes the center of gravity to shift. These changes affect balance and further increase the risk for falls. In addition to a loss of mass in bone cells, muscle mass decreases and is accompanied by decreases in muscle strength and tone.

The attachments known as ligaments and tendons are less elastic, resulting in muscle spasms and decreased flexibility. Pronounced stiffness and diminished range of motion are more noticeable in the morning or following periods of disuse. Complaints of muscle weakness are most frequently caused not by age-related changes but by inactivity.

Cardiovascular System

Normally there is no significant decrease in heart size with advancing age. Heart valves become thicker and more rigid. **Lipofuscin**, a pigmented metabolic waste product, has been found in greater amounts in various organs of the aged body. Loss of elasticity in blood vessels, combined with the accumulation of collagen and lipofuscin, results in narrowing of the **lumen** (diameter), causing a subsequent increase

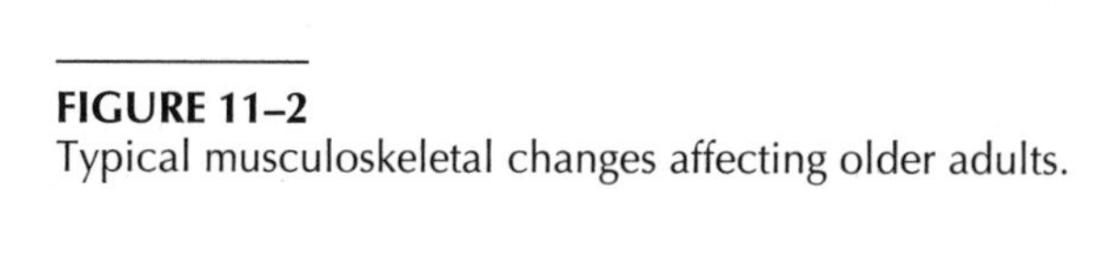

FIGURE 11–2
Typical musculoskeletal changes affecting older adults.

in blood pressure. It is not unusual to have a slight increase in the systolic pressure while the diastolic pressure remains the same. Significant increases in blood pressure are more likely to be the consequence of environmental factors (diet, weight, and stress levels) rather than the result of age. A decrease in cardiac output of 1 percent per year occurs between ages 20 and 80 and is due to the loss of cardiac muscle strength and contractility. This change may be evidenced by a slower heart rate. The older heart needs more rest between beats. Regular exercise can increase cardiac performance and prevent complications. The best type of exercise for maximum cardiac function is walking. Veins become more visible and tortuous. Increased pressure on weak vessel walls leads to the increased incidence of varicosities in the lower extremities and rectum.

Respiratory System

The respiratory system is subjected to a great deal of abuse during the lifetime. The age-related changes are subtle and occur gradually. Several structural changes in the chest diminish respiratory function. Calcification of the rib cage and costal cartilage makes the chest wall more rigid and less compliant. These changes in the thoracic walls make the respiratory muscles work harder. Lung tissue gradually loses elasticity. Vital capacity decreases, and more muscular work is needed to move air in and out of the lungs. Between ages 20 and 60 about 1 L of vital capacity is lost. Lungs exhale less efficiently, causing an increase in the residual volume. The **residual volume** refers to the amount of air remaining in the lungs after forceful exhalation. Coughing is less effective. All of these changes make the older person more susceptible to respiratory infections.

Gastrointestinal System

The numerous changes in the gastrointestinal system cause discomfort but are usually not serious enough to place the person at health risk. There is a decrease in saliva, which results in **xerostomia** (dry mouth) and **dysphagia** (difficulty in swallowing). There is a diminished gag reflex, which places the older person at risk for choking while eating. To decrease the frequency of choking, older persons should eat slowly and in an upright sitting position.

Because of the decrease in **peristalsis**, the muscle movement that propels food through the gastrointestinal system, it takes longer for the esophagus, stomach, and lower intestine to empty. In the esophagus this increases the risk of aspiration. For this reason, older people should not only eat in an upright position but maintain this position for an hour after eating. In the stomach this change, together with decreased gastric secretions, may result in indigestion. There is also a decrease in the total stomach capacity, causing a decrease in hunger and appetite. Changes in the intestines include a decrease in the absorption of nutrients. Individuals who use laxatives on a regular basis may be at further risk for vitamin and nutrient deficiencies. As the liver ages, there is a decrease in enzyme production. This may adversely affect metabolism of both food and drugs.

In the lower intestine, the decrease in peristalsis slows the movement of waste, often producing constipation and increased flatus. In order to maintain normal bowel functioning, older people need to maintain adequate intake of fluids and roughage. Regular toileting habits and exercise will further enhance normal bowel functioning. Bowel movements may also be affected by a decrease in nerve sensations and a delay in the signal to defecate. These changes, along with a weakening of the external sphincter in the rectum, may sometimes cause bowel incontinence.

Dentition

Tooth loss is not a consequence of the aging process but the result of poor care leading to disease. With proper care, older persons can retain their teeth through their entire life. As they age, teeth show natural signs of wear and tear, including a loss of some enamel, lengthening of the tooth, and a decrease in the tooth's ability to cut and chew efficiently. These changes have significant implications for both safety and digestion. Chewing ability and condition of the mouth and teeth should be considered when preparing foods for elderly people. Soft or pureed foods can be substituted for foods of regular consistency if indicated.

Integumentary System

The **integumentary system** consists of the skin, hair, nails, and oil and sweat glands. The skin helps the body maintain a state of **homeostasis** (a state of internal balance). The skin protects the body from changes in temperature, pressure, and moisture and from invading organisms. Normal aging may compromise the skin's ability to maintain homeostasis. The skin will lose some elasticity and become wrinkled. As aging progresses, the skin gets thinner, drier, and more fragile. These changes make the older person more prone to skin breakdown following minor bruising or injury. Normal circulatory changes may delay wound healing in elderly people.

Older people lose subcutaneous adipose tissue, causing a decrease in their ability to sustain changes in temperature. Further complicating their ability to maintain body temperature is a normal decrease in the number and function of the sweat glands. Older persons perspire less and chill easily. They commonly complain of being chilly if they are seated near a window or draft. A sweater or light cover will serve to make the older person feel more comfortable. Elderly people are also more apt to suffer from heat stroke as a result of the changes in their ability to perspire. In warm weather they should avoid overexertion and maintain adequate hydration.

Skin, as it ages, shows irregular pigmentation, **senile lentigo**, due to an uneven distribution of **melanocytes** (pigmented cells). The actual number of melanocytes in the skin decreases as much as 80 percent between ages 27 and 65. Extensive repeated exposure to the sun and ultraviolet rays can further exaggerate the normal aging effects on the skin.

Nail growth gradually slows and the nails become more brittle, with a dull yellow appearance. The toenails may become thicker and should be trimmed by a podiatrist regularly. Fingernails require special attention, including frequent cleaning and trimming.

Hair grays by age 50 in 50 percent of the older population. Hair loss is a common feature beginning in the 30s in men and after menopause in women. Hair loss is not confined to the head but can be seen in other areas, including the axillary and pubic areas. Men show an increase in hair growth in the eyebrows, nose, and ears. Women may note some unwanted hairs on the face and chin.

Many age-related skin changes place elderly people more at risk for skin disorders, such as infections, **pruritus** (itching), **keratosis** (thickening), pressure sores, and skin cancer.

Nervous System

There is a decrease in the number of **neurons**, or nerve cells: 5 to 10 percent of neurons atrophy by age 70, and after age 70 the rate of atrophy increases. The result is a decrease in the nervous system's capacity to transmit messages to and from the brain. Brain weight peaks at age 20, and the brain loses 100 g, or 7 percent of its weight, by age 80. Cerebral blood flow decreases because of changes in the vessels of the circulatory system. However, problems with memory and learning are not the result of these normal changes of aging, but of specific diseases that affect the system's ability to function. Other neural changes include a slowing of motor responses. Reaction time is as much as 30 percent longer in older individuals. Elderly people must be assessed individually to determine their response time and ability to drive safely.

Sensory System

Normal age-related changes in the sensory system may cause problems with daily functioning and general well-being. The five senses—taste, sight, hearing, touch, and smell—all become less efficient, placing the older person at greater risk for injury. The vision changes that begin in middle age continue during this stage. Presbyopia, the loss of eye's ability to focus, and **opacity** (clouding) of the lens progress. The incidence of cataracts and glaucoma increases. Peripheral vision diminishes and sensitivity to glare increases. Color vision changes with aging: red and yellow are seen best, whereas the discrimination of green and blue colors fades. For safety reasons, bright colors like yellow and orange are best used to mark curbs and steps.

Excessive watering of the eyes may result from blockage of the **lacrimal ducts** (tear ducts). Certain medications, vitamins, and diseases can lead to dryness of the eyes. The use of artificial tears helps to alleviate the discomfort and protect the cornea from drying.

About one third of persons over age 65 have sufficient presbycusis (age-related hearing loss) to affect their everyday lives. At age 10 a person can hear frequencies as high as 20 kHz; by age 50 the maximum level is 14 kHz; and by age 60 there is little hearing over 5 kHz. It is best to address older people in low moderate tones to compensate for the loss of high-frequency hearing. Small insults and injuries or certain diseases may contribute to hearing loss. Changes in the middle ear include thickening of the tympanic membrane and calcification of the bones. The accumulation of ear wax (**cerumen**) may interfere with the passage of of sound vibrations through the external canal to the middle and inner ear. Symptoms of this type of hearing loss include feeling of fullness, itching, and **tinnitus** (ringing) in the ears. Conductive hearing may show marked improvement after the removal of accumulated wax. Other types of hearing loss may be related to nerve atrophy and circulatory changes. Box 11–2 offers suggestions for improving verbal communication with the hearing impaired.

BOX 11–2

Improving Verbal Communication for the Hearing-Impaired

- Speak clearly and distinctly in low tones.
- Rephrase words as needed.
- Face the listener.
- Use facial expressions and gestures to help clarify your message.
- Use well-lit areas, placing lighting behind the listener.
- Minimize outside distractions.
- Encourage lip-reading.

The decrease in the number of taste buds causes a loss of taste discrimination, first for sweet and later for other tastes. These changes are not solely age-related; environmental factors may also contribute to the decline in taste. There is little research in the area of smell and the aging process. It is believed that the sense of smell declines with normal aging due in part to olfactory degeneration. Safety is an issue here: older people living alone may not be able to detect subtle gas leaks or smoke. For this reason the use of warning detectors in the home is most important. Significant changes in tactile sensation are thought to be more related to disease than to aging alone.

Genitourinary System

After menopause the ovaries, uterus, and fallopian tubes atrophy. The vaginal walls become thinner and less elastic and there is a decrease in lubrication and vaginal secretions; this may result in discomfort during intercourse. Vaginal secretions normally protect the vagina from bacterial invaders. This protective function diminishes, making older women at greater risk for vaginal infections.

Approximately 2.5 percent of body's calcium may be lost in the first few years after menopause; after this period the rate of bone loss slows down. It is believed that estrogen has an antiatherosclerotic effect, protecting women from heart disease. As estrogen levels decline, the incidence of heart disease in postmenopausal women increases to equal that in men. Other common changes occurring after menopause include deepening of the voice, thinning of the pubic hair, and atrophy of the breast tissue. Hormone replacement therapy may be considered (see Chap. 10).

After age 50 men experience a gradual decline in testicular mass. It takes longer for older men to achieve an erection and less semen is released at ejaculation. Testosterone and sperm levels decrease gradually, but healthy men retain their fertility well into their older years. Hypertrophy of the prostate gland may cause difficulty in voiding. The prostate gland is separated from the rectum by connective tissue, making its posterior surface easily palpable on a digital rectal exam.

Normal age-related changes affecting the urinary system occur gradually. The kidneys decrease in size and lose some of their functional units, or **nephrons**, resulting in a one-third to two-thirds reduction in the filtration rate. This may result in a decreased ability to filter, concentrate, or dilute urine. However, even with these age-related changes, the aging kidneys should continue to maintain homeostasis.

The bladder walls lose elasticity, and bladder volume decreases from 250 to 200 mL. Women with a history of multiple pregnancies are at greater risk for further weakening of the pelvic floor muscles, leading to stress incontinence. The signal indicating a need to void may be delayed, making the older person prone to accidents.

Endocrine System

As individuals age, secretory cells of the endocrine system are replaced with connective tissue, producing a decrease in the hormone levels. As the endocrine system ages, all body tissues and organs are affected by these age-related changes. Diabetes mellitus and thyroid dysfunction are the two main endocrine and metabolic disorders affecting elderly patients.

VITAL SIGNS

Normal age-related cardiovascular changes cause a moderate increase in systolic blood pressure. Hypertension is defined as systolic blood pressure greater than 140 mm Hg with a diastolic of 90 mm Hg or more. Resting heart rate usually remains unchanged or slows slightly. There is little or no change in the older person's resting respiratory rate. However, more muscle work is needed to move air in and out of the lungs.

DEVELOPMENTAL MILESTONES

Motor Development

Changes in both the musculoskeletal and nervous sytems cause movements to slow down with advancing age. Both gross and fine motor skills may be affected by the stiffening of ligaments and joints. Gait speed and step height decrease. Postural and balance changes further affect mobility.

Sexual Development

Contrary to popular belief, elderly people are capable of enjoying a satisfying sexual relationship. Affection and pleasure-seeking behaviors are important and may be expressed differently as individuals age and opportunities present. Men need more stimulation to reach erection. Women can use estrogen creams or other lubrication to prevent discomfort caused by drying of vaginal tissues. Respect for the individual's privacy in all settings helps promote dignity and a positive self-image.

Psychosocial Development

Successful resolution of the first seven stages of Erikson's psychosocial development prepares the older person for the task of **ego integrity**. Ego integrity is similar to a life puzzle in which all the pieces fit nicely together. The task of each stage is to complete the integration of the person, adding meaning to his or her life. Those who develop a sense of ego integrity usually feel satisfied with their accomplishments. They may look back over their lives and admit to certain failures and disappointments, but generally they feel that they have been successful. Ego integrity allows them to proceed with a sense of calm toward death, confident of having left a legacy for future generations.

Wisdom acquired throughout a life of experience is a common characteristic associated with a sense of ego integrity. The process of **reminiscence**, or life review, reassures older people about their accomplishments and worth. Reminiscing allows elderly people to weave their life line together, giving events and memories meaning and order. It facilitates an understanding of the past, puts the past in context, and allows the person to make peace with disappointments and face the future with optimism.

People who feel that their life has no meaning or that they have made the wrong decisions develop a sense of despair. This produces a sense of helplessness and lack of control over their life. A sense of despair is also associated with a fear of death and anxiety about the future.

As with Erikson's other tasks, developing ego integrity is also affected by one's family and other socialization experiences. All of these experiences combine to help shape one's attitude toward aging. If old age is seen as a time of decline and nonproductiveness, these negative expectations are more likely to be fulfilled. **Ageism**, or prejudice against older people, contributes to negative perceptions of the aging process. Educational programs help combat ageism and foster positive attitudes toward older individuals. Furthermore, if young children have good role models, they are more likely to have a positive attitude toward aging throughout their lives, and thus be helped to establish ego integrity themselves.

Achieving ego integrity also involves adjusting to changes in body image, family roles, work and leisure, and sexuality and facing the inevitability of death.

Changes in Body Image

How we view aging will ultimately affect how we cope with the changing body image. Physical appearance has a strong impact on a person's self-concept. Most of the visible changes of aging occur gradually, giving older persons time to adjust to their new image (Fig. 11–3). Individuals whose identity is based solely on their physical attractiveness will continue to see life from that perspective.

Changes in Family Roles

As couples grow old together, they must make several adjustments. Physical or emotional illness in either spouse may frequently be the cause of changes in roles in older marriages. One spouse may become the nurse or caregiver. This may cause anger, resentment, and depression in either the giver or receiver of care. Patterns of dominance may shift from the man to the woman or vice versa. Roles may also change with retirement, placing new limitations and stresses on both parties. Husbands may spend more time at home than ever before, causing conflicts if they try to assume the in-charge role in the home. Other men may find that their role has changed to homemaker while their wives continue to work. Any one of these role changes requires a period of role adjustment.

Death of a spouse produces a change of roles and stress on the remaining individual and other family members. The loss of a spouse is a highly significant life event. Studies indicate that married elderly individuals are generally healthier than unmarried persons. Married persons have a lower incidence of chronic diseases and institutionalization than single, widowed, and divorced individuals. In addition, studies indicate that there is a higher mortality rate among recently widowed men (6 months) and women (2 years). Married women over age 65 are more likely to be the surviving spouse than married men in the same age group.

Two problems common to widowhood are loneliness and decline in income. Frequently the widow finds that she is unprepared to be the decision maker and financial overseer. Former relationships and activities may disappear, forcing her to

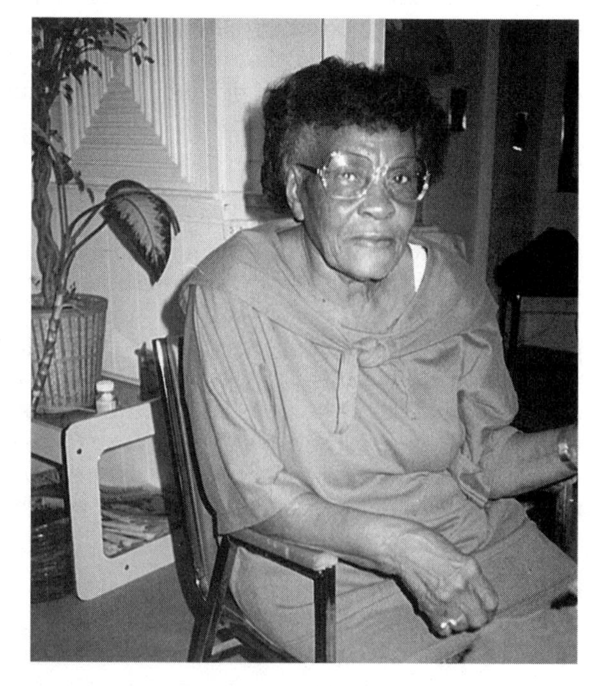

FIGURE 11–3
Older people need to accept their changing body image.

pursue new activities. Given time and support, many women lead independent, well adjusted lives after the death of their spouse. Many elders seek new relationships for companionship. Past marital experiences, good health, adequate income, and the attitudes of grown children are factors important to successful relationships.

Some widowed or divorced elders seek remarriage as an option. For a long time elders were forced to give up their former spouse's Social Security benefits. But Congress has since passed legislation that allows the surviving spouse to choose between the benefits of a former spouse or those of a new spouse, whichever are greater.

Older people may lose friends due to death or relocation. Some older individuals on fixed incomes find it necessary to move to new communities, giving up old friends and neighbors.

The divorce rate of persons over 65 doubled between 1960 and 1979, and this trend has continued. Debilitating illnesses, disabilities, and marital distress are the main reasons cited for divorce in this age group. Anger and guilt both have an adverse effect on widowed and divorced individuals. Divorce also places the individual at risk for economic difficulties. A nursing assessment of the emotional and social support systems available to those who are widowed or divorced can help them through the adjustment period.

Elderly people fear loss of independence more than any other loss, commonly expressing the feeling that they don't want to become a burden to their family. Illness or disability may result in the loss of independence. If this happens, every effort should be made to help the affected individual maximize his or her capabilities and independence for as long as possible. This enhances self-esteem and increases feelings of usefulness.

One role that helps decrease feelings of isolation for many individuals is grandparenting (Fig. 11–4). Most elders have regular frequent contact with their grandchildren. Most children have strong feelings of affection for their grand-

FIGURE 11–4
Grandparenting is a source of satisfaction for both grandparents and grandchildren.

parents. Age and state of health will help to determine the amount of interaction and style of grandparenting they share. Many individuals live long enough to assume the role of great-grandparents.

Changes in Work and Leisure

Work gives many people a sense of identity and self-esteem as well as financial rewards. Many older people continue to work after age 65. In 1994 about 3.8 million older Americans, or 12 percent of the older population, were working or actively seeking work. Older workers place more significance on the social importance of the job, while younger workers usually value income most. Job studies have shown that older employees have a higher rate of job satisfaction, lower absenteeism, and lower job turnover. Even with all of these positive factors, subtle age discrimination still exists against the older worker. Older persons have great difficulty finding and keeping jobs. The job market today continues to show preference toward young workers, hiring them over the older, more skilled people.

Many older adults continue to work in order to postpone retirement, since the discontinuation of one's work role causes a change in lifelong habits. Many people's self-worth is directly tied to their work role. Although they may have dreamed about having more leisure when they were young, retirement may become less appealing as the time approaches. Some men who have been in the workforce for many years may still look forward to retirement. Women, perhaps especially those who entered the workforce late in life, may choose to retire at a later age.

The state of the older adult's health is the most important factor contributing to the adjustment to retirement. Prior work and leisure habits may also influence the adjustment. Factors that help promote positive feelings toward retirement include adequate income, social support, and a strong self-concept. These factors are important in helping today's older adult adjust to the increased length of retirement. Someone who retires today at age 65 may be retired for 20 years or longer. Of course, the best time to plan for retirement is when the individual first enters the workforce. Appropriate planning can help individuals look forward to a peaceful retirement and allow them to continue to grow and be satisfied.

Even with preparation and planning, retirement is a process that involves different feelings and activities. Researchers have identified seven different phases. The *remote phase* is a period of denial with little preparation for the process. During the *near phase* the person may participate in some planning programs. The *honeymoon* period is characterized by a time of euphoria; during this stage people try to do all the activities that they haven't had time for in the past. A sense of *disenchantment* sometimes occurs as reality sets in; individuals may come to terms with their expectations. During the *reorientation phase* individuals must re-establish goals and change their lifestyle. Adjustment to reality is part of the *stability phase*. If the person resumes work or becomes ill or disabled, the retirement role *terminates*.

Although much of our society is work-oriented, many individuals derive satisfaction from the leisure that comes after a productive work life, feeling that they have earned the time to pursue other interests. Leisure, too, can be viewed as "productive" in the sense that it enhances one's sense of well-being and may include activities that benefit others in the community as well. The activities that people pursue vary greatly according to the individual's particular interests.

Changes in Sexuality

Unfortunately, one of the misconceptions still held by our society is that older adults cannot and should not be sexually active. Older people, despite all the myths, continue to have the capacity to enjoy sex well into late life. Sexual expression during this stage not only includes sexual intercourse but touching, cuddling, and masturbation. Sexual behavior fulfills the older person's basic human need for physical closeness, warmth, and intimacy. Studies have shown that sexual behavior patterns of late adulthood correlate with patterns of sexual behavior in earlier years. Some decline in sexual activity may be related to decreases in opportunities or acceptance of the sexless stereotyped image. Sexual difficulties are usually related to poor health, disease, medications, or other social problems.

Often young adults feel that sex is perfectly natural for them but have difficulty seeing their parents' sexuality as natural. Because of these negative attitudes many older people feel guilty about their sexual needs and fear ridicule of their behavior. Some practices in healthcare facilities may further reinforce negative stereotyping. Very few institutions provide privacy and space for couples to be alone or have sex. In fact, some administrators in long-term care settings even go so far as to separate married residents from one another. Caring displays of affection allow people to feel good about themselves. Other benefits of sexual expression include physical exercise and improved circulation for the heart and lungs. For all of these reasons, sexual expression is important and meaningful in the lives of older people. Nurses can foster a climate of support in which sexual expression is not suppressed but accepted as a natural part of life. Box 11–3 lists practices that promote sexual expression in older people.

BOX 11–3

Promoting Sexual Expression in Older People

- Allow for privacy.
- Be nonjudgmental.
- Maximize strengths and minimize weaknesses.
- Encourage attempts at grooming and the use of cosmetics.
- Use clothing that enhances appearance.
- Avoid belittling or ridiculing older people's interest in the opposite sex.

The Inevitability of Death

Until a certain age healthy people usually don't think of the nearness of death. Death takes on different meanings for different people. Some face death with a sense of tranquility while others fear it and feel that their time is running out. Older adults begin to face the reality of dying as their friends and loved ones die. Multiple losses are a part of older people's world. The losses serve to remind them of their own mortality. Loss of a spouse or dear friends may cause loneliness and isolation. Grieving is a process that facilitates the adjustment to the loss. Grief varies in length and severity and is affected by culture and experiences. Prolonged grieving accompanied by intense feeling of anger, guilt, and sadness may be indicative of depression and should be investigated. Community and religious affiliation and support groups can ease the pain and help with adjustment to loss. Spiritual awareness may become strengthened and help guide and support the people through their losses. Death and dying are emotionally difficult issues for many people. Nurses themselves need to be comfortable when working with dying individuals. They must respect the individual's wishes, help the family of a dying person cope with the event, honor living wills, and so on. Although this is a sad time for many people, helping patients in a humane and compassionate manner can ease the process and perhaps make it into something more meaningful and growth-producing for all involved.

Many older people attempt to resolve old conflicts between friends and family members. This helps them move forward with less guilt and unrest. Individuals should be encouraged to openly discuss their feelings about dying and burial plans. Wills designating property and possessions, living wills, and the appointment of a healthcare proxy are needed for peace of mind.

Cognitive Development

Older healthy people may retain their cognitive abilities well into late life. The Wechler Adult Intelligence Scale (WAIS), developed in 1955, continues to be used to assess adults' cognitive ability. This test measures content that is not related to everyday cognitive tasks and therefore is not useful in determining older people's actual performance outside of the testing area. Tests that use material relevant to everyday functioning show that older people have no decline in intelligence with advancing age.

Many factors influence intelligence, including genetic inheritance, education, socioeconomic background, and state of health. Older individuals are generally less competitive, less interested in impressing others with their scores or performance. New material is learned more slowly as one ages. Attitudes toward learning new concepts are different in this age group. The older person may be more reluctant to try new things and to learn. For example, children placed in front of a computer look at it as a challenge to master, while older adults view it as a threat and fear its use. Old stereotypes and beliefs like "you can't teach an old dog new tricks" can become self-fulfilling prophesies. Older individuals usually solve problems by using their life experiences rather than looking for new solutions.

Learning, perception, and cognition may be affected by the normal age-related changes. Older people suffering from sensory losses have more difficulty concentrating on more than one task at a time. They usually find it more difficult to concentrate and eliminate extraneous noise or interference. Reaction time slows down with aging, making it more difficult to process information at a usual rate. Older adults can compensate for these changes and learn more efficiently when they set their own pace.

Memory shows slight changes with advancing age. Most older people remember what is heard better than what is read or seen. Research testing shows that older people are slower to retrieve stored information. Instead of attempting to remember, some elders will say that they "don't know" rather than try to recall. Recent or short-term memory stores a limited amount of information. Remote or long-term memory stores and encodes information in a meaningful mode. Older individuals show greater losses in short-term memory than in long-term memory. They often remember their wedding party better than what they did an hour ago. The state of health and the amount of sensory losses have a great effect on memory. In fact, many individuals who appear to have suffered from marked cognitive losses actually may have profound uncorrected sensory deficits. The use of hearing aids, eyeglasses, and other assistive devices can enhance learning, independence, and self-esteem (Fig. 11–5). Older people suffering from chronic physical illnesses have a greater amount of fatigue, which further reduces their learning abilities. Healthy older adults may choose to return to school or college with very successful outcomes. Those actively pursing their education find satisfaction and self-fulfillment.

Moral Development

The wisdom ascribed to older people since ancient times implies that their level of moral reasoning has reached its optimal point. In reality, however, the moral and ethical concerns of this age group are no different from any other age group. Moral beliefs come with a lifetime of experiences and interactions with others. The older person's basic moral code may change as a result of illness or need. Disease or medication can interfere with the person's moral reasoning. Intact cognitive skills are necessary to utilize moral reasoning.

Many older people have more time to devote to their spiritual needs. There is no evidence, however, that older people become more religious at this stage in their lives. Still, many find meaning to life based on their spiritual beliefs by accepting and following the teachings of a particular religion. Those who have strong beliefs may find peace and satisfaction in their lives.

NUTRITION

Many factors affect older people's nutritional status and eating habits. Nutritional status may be affected by the person's lifestyle, changes in body composition, and use of medications. Inflation and a fixed income may prevent some older people from buying the foods necessary for an adequate diet. Social situation (living alone or

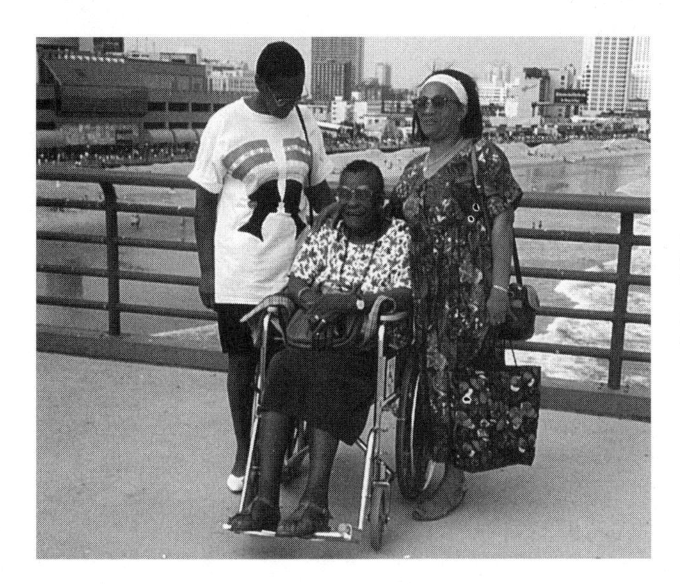

FIGURE 11–5
The use of assistive devices helps maintain independence and self-esteem.

with others) and the level of education may also affect whether older individuals can adequately meet their nutritional needs. Weight maintenance is very important in elderly people. Weight loss in elderly people may reflect actual loss of muscle tissue. Eating habits reflect culture, religion, and family structure. By old age, eating patterns are well-established habits that are difficult to change. A careful review of the diet must be done before attempting a dietary modification. Dietary adequacy is directly affected by income, socioeconomic status, and social situation. Inflation and fixed income further hinder the purchasing power of the older individual. The level of education may also relate to the person's understanding of dietary needs.

Living alone often results in decreased appetite and the desire to prepare meals. Disabilities, sensory losses, and other age-related changes in various body systems may further complicate the older person's ability to shop and to prepare, eat, and enjoy food. Table 11–1 summarizes health implications of diet in elderly

TABLE 11–1
DIETARY IMPLICATIONS FOR OLDER ADULTS

System	Physiological Changes	Dietary Suggestions
Cardiovascular system	Reduced elasticity of blood vessels	Reduce sodium
	Decreased cardiac output	Lose weight if overweight
	Decreased force of contraction	
Respiratory system	Loss of elasticity	Reduce calories to prevent obesity
	Decreased maximum breathing capacity	
Renal system	Decreased blood flow	Drink plenty of fluids
	Decreased filtration, reduced numbers of nephrons	
Neuromuscular system	Decreased responses	
	Decline in physical strength and motor function	
Nervous system	Decline in reaction time	Appetite may be adversely affected by the loss of taste and smell
	Decreased speed of nerve impulses	
Endocrine system	Reduced blood levels of some hormones	Reduce use of simple sugars
	Declining glucose tolerance	
Gastrointestinal system	Loss of teeth	Use dentures
	Decreased taste, saliva, and digestive enzymes	Use broth and juices to moisten foods
	Decreased peristalsis	Eat small, frequent meals
		Increase the amount of fiber
		Maintain adequate fluid intake
		Take vitamin C and adequate protein to aid healing
Skin	Reduced subcutaneous fat	
	Atrophy of sweat glands	
	Skin discolored, thin, wrinkled, and fragile	

people. Other concerns may include inadequate transportation to stores and a fear of violence in the neighborhood. Title VII of the Older Americans Act offers nutritional programs for elderly people. Meals on Wheels is an example of a program used to provide homebound elders with nutritional meals.

Eating patterns by this age are largely fixed by culture, religion, and family structure. If changes in diet are necessary, it is important that the nurse carefully review not only the individual's diet but also all of the factors cited above.

After age 21 the resting basal metabolic rate declines at a rate of 2 percent per decade, resulting in decreased calorie needs. Total calorie needs also depend on the individual's activity level. Between the ages 51 and 75 most men should consume 2000 to 2300 calories per day, while most men age 76 and older need 1650 to 2000 calories per day. Women between ages 51 and 75 usually need 1400 to 1800 calories per day, whereas women aged 76 and older need about 1550 to 1700 calories per day.

Carbohydrates should represent 60 percent of the older person's caloric intake. The best carbohydrates are those that are complex starches and sugars such as whole-grain breads and cereals. These foods are digested over a long period of time and therefore are more satisfying to the appetite. Carbohydrates are also relatively inexpensive, tasty, and capable of being stored for long periods without refrigeration. A carbohydrate intake of 100 g daily is recommended for all adults. If intake goes below 50 g per day there is a danger of developing ketosis, an accumulation of ketones when fats are improperly metabolized. This could lead to a disturbance in the acid-base balance.

Proteins should represent 12 to 14 percent of caloric intake. Because protein foods are more expensive, they may be missing from the older person's diet. A complete range of amino acids are found in eggs, meats, milk, fish, or poultry or in mixtures of rice, beans, cereals, nuts, and seeds. The protein needs of the healthy older adult differ greatly from those of an older ill person. For example, after surgery older individuals need greater amounts of protein to help the body build and repair tissues.

It is important to have some fat daily in the diet to help provide energy, transport fat-soluble vitamins, insulate and cushion the body, and make body compounds. However, it has been suggested that older adults need no more than 30 percent of their daily calories from fats. Diets restricted in cholesterol and high in unsaturated fats have been useful in minimizing the risk of cardiovascular disease.

Dietary fiber has multiple purposes. It helps to prevent constipation by increasing the bulk of the stool. It helps to control weight as it satisfies hunger and gives the sensation of fullness without extra calories. Research shows that it may protect against colon and breast cancer. The recommended level of dietary fiber is 25 to 50 g per day. Foods such as whole grains, brown rice, unpeeled fruits and vegetables, legumes, nuts, and bran all add fiber to the diet. Fluid intake should be increased to prevent the development of fecal impaction and possible intestinal obstruction.

Older people are often less sensitive to the sensation of thirst, may not be able to obtain fluids, or may withhold fluids to prevent nocturia. These factors place older adults at greater risk for dehydration than young adults. Some elderly adults have conditions that further increase their need for fluids, such as diuretic therapy, fever, vomiting, and diarrhea. Normally there should be a balance between water intake and output. There should be sufficient liquid intake to produce 1000 mL of urine daily. Healthcare workers should be alert to the signs of dehydration, which include confusion, dry mouth and tongue, sunken eyes, dry loose skin, a urine specific gravity greater than 1.030, and urine output of less than 500 mL per day.

Many older people take some type of vitamin supplement. Vitamin supplements that meet RDA levels or less are convenient, inexpensive means of providing some dietary needs. They should never be substituted for a balanced diet, however. It is also important that older people be cautioned against using excessive amounts of vitamins.

Alcohol in moderation, one drink a day for women and two drinks a day for men, is associated with lower risks of coronary heart disease in some individuals.

SLEEP AND REST

Older individuals need more rest and less sleep than younger adults. Rest and sleep help to restore the body's energy reserve and prevent fatigue. However, the quality of sleep deteriorates during old age. The older person may take a longer time to fall asleep and may awaken more frequently during the night. It has been reported that 30 to 50 percent of elderly people have difficulty with sleep. Physical discomfort, anxiety, and night-time urination (nocturia) are factors that cause these awakenings. The results of poor sleep are readily apparent the next day, causing signs of poor performance, irritability, and exhaustion.

An assessment of sleep history can be useful in planning nursing interventions. Questions about falling asleep, night waking, and so on, can provide clues to promoting better sleep habits. Nurses should follow individuals' prior sleep habits when caring for them in institutional settings. Many different things can be done to make people comfortable before going to sleep: keeping them warm and covered, offering toileting, and providing a warm drink. Other measures that may help promote sleep include daytime naps, exercise in the early part of the day, avoidance of stimulants (coffee, alcohol, and nicotine), and avoiding large, heavy meals before bedtime. Box 11–4 lists ideas for promoting sleep.

BOX 11–4

Promoting Sleep among Older People

- Meet the comfort needs of the individual (toileting, hygiene, and nutrition).
- Follow the person's normal sleep routine.
- Provide a quiet, relaxing environment.
- Maintain room temperature between 68 and 72°F.

EXERCISE AND LEISURE

Lack of activity results in physical decline in older individuals. Research shows that more than 40 percent of older adults do not engage in exercise or physical activity. Less than one third of this age group participate in moderate physical activity such as walking and gardening. Only 10 percent actually engage in any vigorous activity. Yet exercise has been recognized as a means to help maintain physical fitness across the life span. Regular exercise has been shown to slow the effects of the aging process, maximize the body system's efficiency, and decrease in the incidence of coronary artery disease, hypertension, adult-onset diabetes, colon cancer, anxiety, and depression. Before beginning any exercise program individuals need to consult their physician for medical clearance. Moderation is the key to all exercise programs. Box 11–5 lists the benefits of exercise.

BOX 11–5

Benefits of Exercise

Exercise should be habitual but not unduly strenuous. An example of beneficial exercise is sustained walking for 30 min per day. Regular exercise:

- Reduces the risk for coronary heart disease
- Promotes cardiorespiratory fitness
- Builds muscle strength, endurance, and flexibility
- Is important for weight control
- Lowers blood pressure, blood lipids, and glucose tolerance
- Enhances well-being and helps reduce risk of depression

SAFETY

Decreased auditory and visual acuity, gait changes, and neurological disorders increase the older person's risk of falling. Older people need to be taught safety practices, including getting up slowly and avoiding hot showers, which may make them dizzy. Normal circulatory and skin changes make their skin fragile and more prone to injury. Injuries to the skin heal more slowly.

Statistical studies show that, contrary to popular belief, elderly people are no more likely to be victims of crime than younger adults. They are, however, often victims of purse snatching, thefts, and fraudulent schemes. Elderly persons may limit their activities out of fear that they are easy prey to criminals. This same fear that causes them to stay home may result in some feelings of social isolation and loneliness.

Abuse is defined as the willful infliction of physical or emotional pain or the deprivation of basic care necessary for survival or comfort. The exact number of cases of elder abuse is difficult to estimate, but studies indicate that it is increasing in a variety of settings. Abuse crosses all social, cultural, and socioeconomic boundaries. Most often abuse is related to caregivers' stress, unresolved family conflicts, or families with a history of abuse. All forms of abuse are destructive and at the very least reduce the individual's self-esteem. Healthcare workers must be aware of elder abuse and understand that they are legally required to report acts of abuse that they have witnessed. The individual who fails to report an abusive act may be held responsible by the courts. Box 11–6 lists indicators of elder abuse.

BOX 11–6

Indicators of Elder Abuse

- Evidence of substance abuse
- Social isolation
- Lack of support systems
- Financial problems
- Marital difficulties
- Outward aggression
- Previous psychiatric history, neglect, or mistreatment

HEALTH PROMOTION

The focus for health promotion is different for the older adult than for younger people. The emphasis is no longer solely aimed at prevention but now placed on health maintenance. Exercise, diet modification, and healthy lifestyles can be useful in maximizing wellness and reducing risk factors in this age group. Health education and positive attitudes toward aging also help promote health during the later years. Lack of knowledge about health and health promotion and the devaluation of old age can prevent older people from seeking proper healthcare services.

Health screening and maintenance include tests to detect for abnormalities and illnesses at early stages. All older adults need an annual physical examination. Health examinations should include an assessment of diet, activity level, medication usage (prescribed and over-the-counter), smoking, and alcohol intake. Vision and hearing tests should also be conducted annually. Immunizations are important at this time as there is a decrease in the immune response. Vaccinations against tetanus, pneumonia, and influenza are important. Pneumococcal pneumonia is three times more common in people 65 and older. In 1987 pneumonia was responsible for an average of 48 days of decreased activity per 100 people aged 65 years and older.

Elderly people are more likely to suffer from at least one chronic condition and many have multiple conditions. In 1993 the most frequently occurring conditions per 100 elderly were arthritis, hypertension, and heart disease. Older people accounted for 36 percent of all hospital stays. Elderly patients had an average length of stay of 7.8 days, as compared to 4.9 days in the under-56 age group. Older people use a greater number of healthcare services than younger people and account for 36 percent of total personal healthcare expenditures. This translates to four times the amount of healthcare expenses of younger individuals.

The leading causes of death among elderly people are heart disease, cancer, stroke, arteriosclerosis, diabetes, lung disease, and cirrhosis of the liver. Breast cancer is a concern after age 50. Early diagnosis has proven to be effective in increasing survival rates. Mortality rates have been reduced 25 to 35 percent in women ages 50 to 59 with proper mammography screening.

Older people are particularly susceptible to ultraviolet rays because of their normal age-related skin changes. Brief periods of sunlight may precipitate photosensitive reactions in older adults who are taking certain medications. Older clients should be taught measures that promote healthy skin, including adequate fluid intake and avoidance of skin and perfume products that contain alcohol. They should be instructed to use an emollient after bathing to help keep the skin soft and moist.

Older individuals today use a wide range of services in an ever-changing healthcare arena. For acute or serious illnesses they usually require inpatient hospitalization. Long-term care facilities provide elderly residents with skilled nursing services and rehabilitation as needed. With increasing numbers of older patients being discharged from hospitals after brief stays, the level of acuity in long-term care facilities has increased dramatically. Box 11–7 lists factors to consider when selecting a nursing home. Nurses practicing in these settings now need to expand their understanding and knowledge of the more acutely ill elderly resident. Other older individuals have opted for community-based services such as home care. The typical older home-care client has multiple complex healthcare problems, challenging the skills of home-care nurses. Today's home-care nurse, in addition to possessing excellent clinical skills, must be a self-directed member of a multidisciplinary team. Hospice services provide support and nursing care in the home and in patient-care settings to older people with terminal illnesses. Mental health services are offered in both the community and hospital settings. These services provide health care and maintain psychological well-being. A variety of outpatient services are available to older adults: senior centers, day care, and respite care for clients with Alzheimer's disease. Older people who are homebound can benefit from homemaker services and visiting nursing care.

BOX 11–7

Factors to Consider in Selecting a Nursing Home

- *Costs:* daily rate, insurance accepted, services available
- *Administration:* ownership, accessibility of medical services
- *Philosophy of care:* staff selection, accessibility of staff, approach to residents
- *Other services:* speech therapy, physical therapy, social services, occupational therapy

There are a number of special health concerns and health-related issues for elderly people, including dementia, depression and suicide, and government programs that assist and protect older people.

Dementia

Alzheimer's disease is the leading cause of dementia and cognitive impairment affecting the older age group. There are other similar but treatable forms of dementia characterized by loss of memory, disorientation, and poor social functioning. Among people 65 and older, the incidence of dementia is 5 to 10 percent; for those over 80 it is 20 to 40 percent. Ten to twenty percent of these cases are caused by drug toxicity. The age-related physiological changes affecting the liver and kidneys increase the incidence of drug toxicity. Drug metabolism and excretion slow with aging.

Depression and Suicide

Multiple losses, disease, and medication may lead to depression in the older age group. Again, stereotypical beliefs about older people often prevent families and health professionals from properly identifying depression in elderly persons. Symptoms include feelings of hopelessness and profound feelings of sadness. The use of support services may help cushion the many losses that are experienced at this age. Proper diagnosis and treatment of disease may help prevent depression in some cases. In chronic conditions with debilitation and unrelenting pain, depression is common. Some cases of depression are related to the medications that are prescribed for older people. Polypharmacy, or too many medications, can create pseudodepression. Reducing or eliminating the medication may help reverse the symptoms. Once diagnosed, depression often responds favorably to medication and other treatment modalities.

Untreated depression may lead to suicide. The incidence of suicide in elderly persons is rapidly increasing. Family members and caregivers must be alert to sudden changes in mood or other possible warning signs of suicide. See Chapter 9 for these signs.

Social Security

The Social Security Act was established in 1935 as part of President Roosevelt's New Deal. The original intent of Social Security was to supplement income after retirement. The system is comprehensive and jointly administered by the state and federal governments. Funding for Social Security benefits is from payroll taxes deducted from both employer and employee. On retirement, workers receive benefits equal to their contributions over their working years.

In 1939 the Social Security Act was strengthened to provide millions of older Americans with assistance and as a means toward a better standard of living. Benefits were recalculated based on average earnings over a shorter period of time rather than over a lifetime. Another amendment in 1950 provided benefits to surviving spouses and dependents. It later included government employees and self-employed workers, who were originally excluded. In 1957 disability insurance was set up to provide funds to workers over age 50 who became disabled on the job. In 1960 this was expanded further to remove the age barrier and offer benefits to any disabled worker.

Social Security benefits have tried to keep pace with inflation and rising healthcare costs. At this time benefits have increased faster than earnings; the Social Security Administration has projected that at the present rate funds may be depleted by the year 2015. This is the time when the "baby boomers" of the 1950s will become eligible for pensions.

Medicare and Medicaid

Medicare and Medicaid are two programs that provide universal healthcare coverage to the older population. Medicare accounts for about 45 percent of the funding used by elderly people for healthcare costs, while Medicaid accounts for 12 percent. The remaining costs are often paid by private insurance or by the individual. Medicare, which comes under Title XVIII of the Social Security Act and provides insurance for the aged or disabled, covers acute healthcare services, including physician, hospitals and nursing homes, and home-care services. In 1992, 96 percent of all elderly people were receiving Medicare coverage. Medicare is one of the most expensive federal programs in operation. Medicare is divided into two parts. Part A is financed by mandatory contributions from both employer and employee and functions in similar fashion to hospital insurance. Part B is a voluntary supplemental medical insurance financed through premiums and general revenues. Premiums are about $42.50 per month, with a $100 deductible. This pays for doctor fees and other services. Recipients must be 65 years of age or older or disabled and entitled to Social Security benefits.

Medicare recipients are now encouraged to enroll in managed-care programs rather than to using the fee-for-service method. As of 1966 3.7 million people were enrolled in such services.

Medicaid comes under Title XIX of the Social Security Act and was first introduced in 1965. It is financed jointly by the federal and state governments and provides health care for needy people of all ages. Medicaid is the main source of financing for long-term care. Even with these programs, many people cannot get the healthcare services they need because out-of-pocket costs are often beyond their means. To be fully entitled to "free care," they would have to surrender all of their resources. Private insurance companies have set up a Medigap program. This coverage pays for charges not covered by Medicare.

Rights of Elderly People

In 1987 the federal Older Americans Act was established. This act is designed to protect institutionalized elderly people by means of ombudsman programs. The ombudsman acts as a representative and spokesperson for older clients, making certain that their rights are protected.

In 1990 Congress passed the Patient Self-Determination Act, which was intended to ensure that patients' wishes would be followed if they were unable to speak for themselves. This is accomplished by using one of two types of advance directives: the living will and the healthcare proxy. Living wills are written when the individual is still competent and able to determine the type of future treatment they desire. The healthcare proxy, or durable power of attorney, designates someone to make decisions in the event that the person is unable to do so.

Do Not Resuscitate (DNR) orders can be written by a physician at a patient's or family's request. This document legally protects both doctors and heathcare workers in the event of a sudden death. It is imperative that nurses become familiar with the laws in their state, since all states have varying laws regarding living wills, DNR orders, and other such advance directives.

SUMMARY

1. Old age is divided into three periods: young old (65 to 74), old (75 to 90), and very old (90 and over).
2. Elderly people are the fastest-growing segment of the population in this country. Life expectancy is longer for women than for men. The

single most effective predictor of life expectancy is one's biological parents.

3. Several theories exist and attempt to explain aging. Biological theories include the clockwork theory, free-radical theory, wear-and-tear theory, immune-system-failure theory, and autoimmune theory.
4. Several psychosocial theories attempt to explain how aging affects socialization and life satisfaction: disengagement theory, activity theory, and continuity-developmental theory.
5. Many physical changes occur as a part of the normal aging process.
6. Psychosocial tasks for old age include accepting and adjusting to changing body image, family roles, work and leisure patterns, and sexuality, and facing the inevitability of death.
7. According to Erikson, the older individual who has accomplished the first seven developmental tasks can now set out to achieve the task of ego integrity. People who lack ego integrity develop a sense of helplessness and despair.
8. Older people use the process of reminiscing, or life review, to help give meaning to their lives and reinforce their worth.
9. Elderly people usually retain their cognitive abilities until late in life. Memory shows slight changes with advancing age. Older people tend to show greater losses in short-term memory than in long-term memory.
10. Moral beliefs come from a lifetime of experiences and interactions with others. Many elders find peace and satisfaction through spirituality and religion.
11. Good nutrition has been shown to prevent late-life diseases and to improve the person's response to treatment.
12. Older individuals need more rest and less sleep than younger adults. Rest and sleep help to restore the body's energy reserve and prevent fatigue.
13. Exercise has been recognized as a means of maintaining physical fitness across the life span.
14. Older people are more likely to suffer from at least one chronic condition; many have multiple conditions.
15. The leading causes of death among elderly people are heart disease, cancer, strokes, arteriosclerosis, diabetes, lung disease, and cirrhosis of the liver.
16. The focus of health promotion and health maintenance is on exercise, diet modification, and healthy lifestyle.
17. Accidents can be prevented by recognizing the increased risk factors unique to this age group. Changes in sensory perceptions, gait changes, and neurological disorders may increase the older person's risk for falls.
18. Crime and abuse with elderly people as victims cross all social, cultural, and economic boundaries.
19. Social Security, Medicare, and Medicaid are programs that provide assistance to older people.
20. Depression and suicide among elderly persons may be the result of multiple losses, diseases, and medication usage.

CRITICAL THINKING

Alice Thompson, an 80-year-old woman, lives with her daughter and son-in-law. She is beginning her second week as a day-care client in a long-term care facility. She has been very quiet and withdrawn since her arrival. Today she responds to your interaction by stating, "My daughter is not human." She asks you to promise to tell her daughter that she wants her Social Security check back.

1. What would be your first verbal response to this client?
2. Outline a planned intervention for Alice and her family.

Multiple-Choice Questions

1. The majority of older Americans live in:
   a. Chronic long-term facilities
   b. Acute rehabilitation facilities
   c. Hospitals as patients
   d. Their own home alone or with their family

2. The psychological theories of aging serve to explain:
   a. Physical changes of aging
   b. A person's life satisfaction
   c. Life expectancy
   d. Life span

3. Older people are at increased risk of falling because of:
   a. An accumulation of cerumen
   b. Postmenopausal symptoms
   c. A shift in their center of gravity
   d. Marked decrease in height

4. As the older person loses adipose tissue:
   a. Muscle weight increases
   b. Memory loss increases
   c. Temperature control is difficult
   d. Their biological clock speeds up

5. The characteristic hearing loss of old age means that the nurse must communicate in:
   a. Low, moderate tones
   b. High-frequency tones
   c. A soft whisper
   d. A loud shout

6. The following retirement phase is characterized by a feeling of euphoria:
   a. Remote
   b. Reorientation
   c. Stability
   d. Honeymoon

Suggested Readings

Anderson, M, and Braun, J: Caring for the Elderly Client. FA Davis, Philadelphia, 1995.

Appling, S: Wellness Promotion and the Elderly. Med Surg Nurs 6(1):45–46, 1997.

Beckerman, A, and Nirthrop, C: Hope, chronic illness and the elderly. J Gerontol Nurs 22(5):19–25, 1996.

Belsky, J: The Psychology of Aging: Theory, Research, and Interventions. Brooks/Cole, Pacific Grove, CA 1990.

Campbell, W: Dietary protein requirements of older people: Is the RDA adequate? Nutrition Today 31(5):192–197, 1996.

Cavanaugh, J: Adult Development and Aging. Wadsworth, Belmont, CA, 1990.

Cutillo-Schmitter, TA: Aging: Broadening our view for improved nursing care. J Gerontol Nurs 22(7):31–42, 1996.

Eliopoulos, C: Gerontol Nursing. JB Lippincott, Philadelphia, 1993.

Fishman, P: Healthy people 2000: What progress toward better nutrition? Geriatrics 51(4):38–42, 1996.

Hofeland, S: Sexual dysfunction in the menopausal woman: Hormonal causes and management issues. Geriatr Nurs 17(4):161–165, 1996.

Holm, K: Why women should exercise. Reflections 22(3):11–12, 1996.

Grant, L: Effects of ageism on individual and healthcare providers' response to healthy aging. Health & Social Work 21(1):9–15, 1996.

Kane, R, Ouslander, J, and Abrass, I: Essentials of Clinical Geriatrics. McGraw-Hill, New York, 1994.

Kart, C: The Realities of Aging: An Introduction to Gerontology. Allyn & Bacon, Boston, 1990.

Kessenich, C, and Rosen, C: Osteoporosis: Implications for elderly men. Geriatr Nurs 17(4):171–174, 1996.

Lanza, M: Divorce experienced as an older woman. Geriatr Nurs 17(4):166–170, 1996.

Larue, G: Geroethics. Prometheus Books, Buffalo, NY, 1992.

Letvak, S, and Schoder, D: Sexually transmitted diseases in the elderly: What you need to know. Geriatr Nurs 17(4):156–160, 1996.

Lutz, C, and Przytulski, K: Nutrition and Diet Therapy. FA Davis, Philadelphia, 1994.

Matras, J: Dependency, Obligations, and Entitlements: A New Sociology of Aging, the Life Course, and the Elderly. Prentice-Hall, Englewood Cliffs, NJ, 1990.

Miller, A, and Champion, V: Mammography in older women: One-time and three-year adherence to guidelines. Nurs Res 45(4):239–235, 1996.

Melilo, K: Medicare and Medicaid: Similarities and differences. J Gerontol Nurs 22(7):12–21, 1996.

Miller, C: Summer skin precautions to protect against adverse medication effects. Geriatr Nurs 17:193–194, 1996.

Nickols-Richardson, S, Johnson, MA, Poon, L, and Martin, P: Demographic predictors of nutritional risk in elderly people. J App Gerontol 15(3):361–375, 1996.

Noe, C, and Barry, P: Healthy aging: Guidelines for cancer screening and immunizations. Geriatrics 51(1):75–84, 1996.

Palmore, E: Ageism: Negative and Positive. Springer, New York, 1990.

Santrock, J: Life-Span Development. Brown & Benchmark, Madison, WI, 1995.

Stanley, M, and Beare, P: Gerontological Nursing. FA Davis, Philadelphia, 1995.

Stocker, S: Six tips for caring for aging parents. Am J Nurs 96(9):32–33, 1996.

Tunstull, P, and Henry, M: Approaches to resident sexuality. J Gerontol Nurs 22(6):37–42, 1996.

U.S. Department of Health and Human Services: Healthy People 2000: National Promotion and Disease Prevention Objectives. DHHS Publication No. (PHS) 91-50212, 1992.

Waldo, M, Ide, B, and Thomas, P: Postcardiac-event elderly: Effect of exercise on cardiopulmonary function. J Gerontol Nurs 21(2):12–19, 1995.

Yen, P: Eating is a pleasure. Geriatr Nurs 17(4):191–192, 1996.

Appendix A
Community Help Services

Abuse Hotline
1-800-621-HOPE

Alcoholics Anonymous
75 Riverside Drive
New York, NY 10115
(212) 870-3400

American Association of Retired Persons (AARP)
601 E Street NW
Washington, DC 20049
(202) 434-2277

American Heart Association
7320 Greenville Avenue
Dallas, TX 75231
(214) 373-6300

American Psychiatric Association
1400 K Street NW
Washington, DC 20005
(202) 682-6000

Child Help USA Inc.
6463 Independence Avenue
Woodland Hills, CA 91370
(800) 4-A-CHILD

Meals on Wheels Foundation
4101 Nebraska Avenue NW
Washington, DC
(202) 966-8111

National Adoption Center
1218 Chestnut Street
Philadelphia, PA 19107
(800) TO-ADOPT

National Adoption Hotline
Washington, DC 20009-6207
(202) 328-8072

National Committee for the Prevention of Child Abuse
332 S Michigan Avenue
Chicago, IL 60604-4357
(800) 244-5373

National Council on Alcoholism and Drug Dependence
12 West 21st Street
New York, NY 1001
(800) NCA-CALL

National Council on Child Abuse and Family Violence
115 Connecticut Avenue NW
Washington, DC 20036
(800) 222-2000

National Council on Elder Abuse
American Public Welfare Association
810 First Street NE
Washington, DC 20002-4267
(202) 635-8985

National Domestic Violence Hotline
(800) 799-7233

National Institute on Aging
31 Center Drive
9000 Rockville Pike
Bethesda, MD 20892-2292
(301) 496-1752

Parents without Partners
8807 Colesville Road
Silver Springs, MD 20910
(301) 588-9354

Planned Parenthood Federation of America
810 Seventh Avenue
New York, NY 10019
(800) 829-PPFA

Appendix B
Recommendations for Health Promotion

AGE	IMMUNIZATIONS	SCHEDULE	HEALTH PROMOTION	RECOMMENDED SCREENING
Birth to 18 months	Diphtheria-tetanus-pertussis (DTP) vaccine Oral poliovirus vaccine (OPV) Measles-mumps-rubella (MMR) vaccine Haemophilus influenzae type b (HIB) conjugate vaccine Varicella (chicken pox)	2, 4, 6, 15, and 18 months 12–18 months	Bottle-mouth syndrome Ocular malalignment Signs of child abuse or neglect Use of proper car restraints Use of smoke detectors Stair and window guards Safe storage of poisons Syrup of ipecac on hand	Height and weight Hemoglobin and hematocrit Phenylalanine Ophthalmic antibiotics Hg electrophoresis T4/TSH Hearing
Ages 2 to 4	Diphtheria-tetanus-pertussis (DTP) vaccine Oral polio virus vaccine (OPV)	Between ages 4 and 6 years	Safety restraints Smoke detectors Window guards Bicycle safety helmets Safe storage of poisons, matches, and firearms Toothbrushing and visits to the dentist Dietary supervision and exercise	Height and weight Blood pressure Eye exam for amblyopia and strabismus Urinalysis Tuberculin skin test (PPD) Hearing
Ages 7 to 12	Tuberculin skin test (PPD) Varicella (chicken pox)	11 to 12 years if not given earlier	Safety belts Smoke detectors Bicycle safety helmets Safe storage of poisons, firearms Regular toothbrushing and dental visits Dietary supervision and exercise	Height and weight Blood pressure Tuberculin skin test (PPD)
Ages 19 to 39	Tetanus-diphtheria (TD) booster	Every 10 years	Avoidance of substance abuse Safe-sex practices Use of seat belts Smoke detectors Regular toothbrushing and dental visits Dietary education and exercise	Dietary intake Physical activity Tobacco/alcohol/drug use Sexual practices
Ages 40 to 64	Tetanus-diphtheria (TD) booster	Every 10 years	Avoidance of substance abuse Safe-sex practices Use of seat belts Smoke detectors Regular toothbrushing and dental visits	Dietary intake Physical activity Tobacco/alcohol/drug use Sexual practices
Ages 65 and over	Tetanus-diphtheria (TD) booster Influenza vaccine Pneumococcal vaccine	Every 10 years Annually Anually	Avoidance of substance abuse Safe-sex practices Glaucoma testing Use of seat belts Smoke detectors Regular toothbrushing and dental visits	Symptoms of transient ischemic attack Dietary intake Physical intake Tobacco/alcohol/drug use Functional status at home

Appendix C
Sample of Living Will and Designation of Health Care Surrogate*

INSTRUCTIONS

FLORIDA LIVING WILL

PRINT THE DATE

Declaration made this ______ day of ________________________, 19_____.

PRINT YOUR NAME

I, ____, willfully and voluntarily make known my desire that my dying not be artificially prolonged under the circumstances set forth below, and I do hereby declare:

If at any time I have a terminal condition and if my attending or treating physician and another consulting physician have determined that there is no medical probability of my recovery from such condition, I direct that life-prolonging procedures be withheld or withdrawn when the application of such procedures would serve only to prolong artificially the process of dying, and that I be permitted to die naturally with only the administration of medication or the performance of any medical procedure deemed necessary to provide me with comfort care or to alleviate pain.

It is my intention that this declaration be honored by my family and physician as the final expression of my legal right to refuse medical or surgical treatment and to accept the consequences for such refusal.

In the event that I have been determined to be unable to provide express and informed consent regarding the withholding, withdrawal, or continuation of life-prolonging procedures, I wish to designate, as my surrogate to carry out the provisions of this declaration:

PRINT THE NAME, HOME ADDRESS AND TELEPHONE NUMBER OF YOUR SURROGATE

Name: ____

Address: ____

____ Zip Code: ______________

Phone: ____

*Reprinted by permission of Choice in Dying, Inc., 200 Varick Street, New York, NY 10014. (212) 366-5540.

PRINT NAME, HOME ADDRESS AND TELEPHONE NUMBER OF YOUR ALTERNATE SURROGATE

I wish to designate the following person as my alternate surrogate, to carry out the provisions of this declaration should my surrogate be unwilling or unable to act on my behalf:

Name: ____________________

Address: ____________________

____________________ Zip Code: __________

Phone: ____________________

ADD PERSONAL INSTRUCTIONS (IF ANY)

Additional instructions (optional):

I understand the full import of this declaration, and I am emotionally and mentally competent to make this declaration.

SIGN THE DOCUMENT

Signed: ____________________

WITNESSING PROCEDURE

TWO WITNESSES MUST SIGN AND PRINT THEIR ADDRESSES

Witness 1:

Signed: ____________________

Address: ____________________

Witness 2:

Signed: ____________________

Address: ____________________

Courtesy of Choice In Dying, Inc. 6/96
200 Varick Street, New York, NY 10014 212-366-5540

FLORIDA DESIGNATION OF HEALTH CARE SURROGATE

INSTRUCTIONS

PRINT YOUR NAME

Name: ____
(Last) *(First)* *(Middle Initial)*

In the event that I have been determined to be incapacitated to provide informed consent for medical treatment and surgical and diagnostic procedures, I wish to designate as my surrogate for health care decisions:

PRINT THE NAME, HOME ADDRESS AND TELEPHONE NUMBER OF YOUR SURROGATE

Name: ____

Address: ____

______________________________ Zip Code: ____________

Phone: ____

If my surrogate is unwilling or unable to perform his duties, I wish to designate as my alternate surrogate:

PRINT THE NAME, HOME ADDRESS AND TELEPHONE NUMBER OF YOUR ALTERNATE SURROGATE

Name: ____

Address: ____

______________________________ Zip Code: ____________

Phone: ____

I fully understand that this designation will permit my designee to make health care decisions and to provide, withhold, or withdraw consent on my behalf; to apply for public benefits to defray the cost of health care; and to authorize my admission to or transfer from a health care facility.

ADD PERSONAL INSTRUCTIONS (IF ANY)

Additional instructions (optional):

I further affirm that this designation is not being made as a condition of treatment or admission to a health care facility. I will notify and send a copy of this document to the following persons other than my surrogate, so they may know who my surrogate is:

PRINT THE NAMES AND ADDRESSES OF THOSE WHO YOU WANT TO KEEP COPIES OF THIS DOCUMENT

Name: ______________________________

Address: ______________________________

Name: ______________________________

Address: ______________________________

SIGN AND DATE THE DOCUMENT

Signed: ______________________________

Date: ______________________________

WITNESSING PROCEDURE

TWO WITNESSES MUST SIGN AND PRINT THEIR ADDRESSES

Witness 1:

Signed: ______________________________

Address: ______________________________

Witness 2:

Signed: ______________________________

Address: ______________________________

Courtesy of Choice In Dying, Inc. 6/96
200 Varick Street, New York, NY 10014 212-366-5540

Appendix D
Answers to Multiple-Choice Questions

Chapter 1
1. b
2. c
3. c

Chapter 2
1. a
2. a
3. d
4. c

Chapter 3
1. c
2. c
3. b
4. d

Chapter 4
1. b
2. c
3. b
4. a
5. a
6. d
7. c

Chapter 5
1. b
2. b
3. b
4. b
5. d
6. b

Chapter 6
1. c
2. c
3. b
4. b
5. c
6. d

Chapter 7
1. c
2. d
3. c
4. b
5. b
6. b

Chapter 8
1. b
2. c
3. d
4. b
5. b
6. c

Chapter 9
1. d
2. c
3. c
4. b
5. b
6. a

Chapter 10
1. b
2. d
3. c
4. d
5. c
6. d

Chapter 11
1. d
2. b
3. c
4. c
5. a
6. d

Glossary

acrocyanosis a bluish discoloration of the newborn's hands and feet as a result of poor peripheral circulation.
activity theory the aging theory that suggests that individuals remain active and engaged throughout their later years.
acuity sharpness or clearness.
adenohypophysis the anterior lobe of the pituitary gland.
adolescence a transitional period beginning with sexual maturity and ending with growth cessation and movement toward emotional maturity.
adducted referring to movement of the extremities toward the center of the body
aerobic exercise exercise that works the large body muscles and elevates the cardiac output and metabolic rate.
ageism discrimination against older persons.
ambivalence an emotional state of having conflicting, opposite feelings, such as love and hate for a person or object.
amblyopia known as "lazy eye," a condition seen in early childhood caused by weaker muscles in one eye that, if not corrected, may lead to blindness.
anorexia nervosa an eating disorder that is characterized by willful starvation and severe weight loss.
antioxidants agents that prevent the formation of free radicals and may affect the aging process.
apathy a lack of interest in the surroundings.
Apgar score an assessment scale used to indicate an overall picture of the newborn's status.
apnea an absence of respirations.
apocrine glands sweat glands in the axillae and pubic region.
atrophy wasting away.
autoimmune theory a theory of aging which suggests that aging is related to the body's weakening immune system, which fails to recognize its own tissues and may destroy itself.
autonomy independence and a sense of self.
basal metabolic rate the amount of energy that a individual uses at rest.
benign prostatic hypertrophy a benign enlargement of the prostate gland which causes difficulty voiding, diminished urinary stream, dribbling, and frequency.
blastocyst the developing mass of cells at the point of implantation.
bottle-mouth syndrome dental caries caused by sugar in the milk or juice usually in the nighttime bottle.
bulimia an eating disorder characterized by a series of binges followed by purging or self-induced vomiting.
carcinogens cancer-producing agents such as cigarettes, radiation, etc.
cataract a cloudy formation on the lens of the eye.
cephalocaudal a directional term that refers to growth and development that begins at the head and progresses downward towards the feet.
cerumen wax buildup found in the ear.
cervix the lower "neck" portion of the uterus.
cholesterol a component of many foods in our diet. It is an essential component of cells in the brain, nerves, blood, and hormones.
chromosomes substances that carry the genes that transmit the inherited characteristics.
circumcision surgical removal of the foreskin performed for hygienic or religious reasons.
cleft palate incomplete congenital formation and nonunion of the hard palate.
climacteric change of life.
clockwork theory a theory of aging suggesting that connective-tissue cells have an internal clock that determines length of life.
coitus another term for sexual intercourse.
colostrum precursor of breast milk, present as early as the 7th month of fetal life.
compatibility describes sympathetic, comfortable feeling tone in a relationship.
conception union of the female ovum and the male sperm cell, also called fertilization.
conscience a person's internal system of values, similar to the superego.
continuity-developmental theory an aging theory that suggests that aging should be viewed as a part of the life cycle, not as a separate terminal stage.
cooperative/associative play play style typical of preschool children in which they begin to take turns and share in a cooperative manner.
culture all of the learned patterns of behavior passed down through generations.

deciduous teeth "baby" or primary teeth which usually appear about 6 to 7 months of age.
demographics the study of a group of people including the size of the group, the changes within the group, and the information about where the group lives.
dental caries tooth decay.
depression prolonged feelings of profound sadness and unworthiness.
dermis the inner layer of the skin directly below the epidermis.
development the progressive acquisition of skills and the capacity to function.
dilation widening or expansion of an opening.
disease prevention divided into three levels, primary, secondary, and tertiary. Aimed at disease prevention, includes proper education, nutrition, exercise, and immunization. Early diagnosis and treatment help to prevent permanent disability. When permanent disability arises the aim is toward maximizing level of functioning.
disengagement theory a theory of aging that suggests that older persons and society gradually withdraw from one another to assist the transfer of power from the old to the young.
dominant genes those genes that are more capable of expressing their trait over other genes.
dysfunctional family a family that is unable to offers its members a stable structure.
dyspareunia pain or discomfort during intercourse.
dysphagia difficulty swallowing.
effacement shortening and thinning of the cervix.
ego the executive of the mind. It relates most closely to reality.
ego integrity a period of self-satisfaction that occurs during old age.
egocentricity self-centered thought or actions.
ejaculation the release of sperm and semen.
Electra complex a young girl's sexual attraction for her father and unconscious wish to replace her mother.
embryo the developing organism until the end of the eighth week.
emotions expressed feeling tones which influence a person's behavior.
empowerment a form of self-responsibility that demands that people take charge of their own decision making.
empty-nest syndrome a term used to define the reaction to having grown children leave the home.
engrossment the process of neonatal-father bonding.
enuresis bedwetting after the age of normal urinary control.
epiphyseal cartilage the center for ossification and growth at the end of long bones.
equilibrium a balance or a state of homeostasis.
estrogen hormone produced by the ovary.
eustachian tube the structure that connects the pharynx to the middle ear.
fertilization the union of the female ovum and male sperm cell, also called conception.
fetus the developing organism from the eighth week on.
fibrocystic breast disease benign cystic growths found in breast tissue.
fight-or-flight response a state of readiness to attack or flee.
fontanels commonly called soft spots, spaces found between the infant's cranial bones where the sutures cross.
free radical theory a theory of aging involving highly unstable molecules.
free radicals chemical substances produced by metabolism and thought to play a role in cellular aging.
functional family a family that fosters the growth and development of its members.
general adaptation syndrome (GAS) a response to stress that was described by Hans Selye.
generativity Erikson's task for middle-aged adults. It involves individuals' desire to serve the larger community and have a positive influence on their children.
genes found on strands of deoxyribonucleic acid or DNA within the cell nucleus.
gerontology the study of the normal aging process.
gingivitis an inflammation of the gums characterized by swelling, redness, and bleeding.
glaucoma a disease of the eye characterized by an increased intraocular pressure.
gonads the male and female sex glands.
growth an increase in physical size.
health a state of complete physical, mental, and social well-being, not merely the absence of disease or infirmity.
health promotion health care directed toward increasing one's optimum level of wellness.

health restoration that which begins once the disease process is stabilized.
heredity all characteristics that are transmitted through the genes and determined at the time of fertilization.
holistic including not only physical aspects of health but also psychological, social, cognitive, and environmental influences.
homeostasis a balance.
hormone replacement therapy (HRT) a treatment for menopause characterized by estrogen replacement therapy.
hot flashes sensation caused by vasodilatation of the capillaries and produce a rush of blood to the skin surface.
hypertension the term that refers to high blood pressure.
id the body's basic, primitive urges.
immune-system-failure theory a theory of aging that suggests that aging results when the immune system is unable to perform its function.
industry Erikson's task for the school-age child. At this stage children are productive and focus on the real world and their part in it.
infant mortality rate the number of infant deaths before the first birthday per 1000 live births.
initiative Erikson's task for the preschool child demonstrated by pretend, exploration, and the trying on of new roles.
insomnia inability to sleep.
integumentary system skin and related structures.
intimacy a feeling that involves warmth, love, and affection.
introspection a form of self-reflection, may serve as a tool that permits sharing of our innermost thoughts.
involution the return of the uterus to its nonpregnant state.
keratosis a skin thickening.
karyotype the individual chromosomal pattern of a person.
kyphosis curvature of the thoracic spine.
lacrimal ducts tear ducts.
lanugo soft, fine, downy hair covering newborns.
larynx a term for the voice box which houses the vocal cords.
latency the period described by Freud when school-age children's sexual energies are relatively dormant.
libido sex drive.
life expectancy the average number of years a person is likely to live.
life span the maximum number of years that a species is capable of surviving.
lifestyle a person's habits and usual practices common to daily living.
lipofuscin a pigmented metabolic waste product that has been found in greater amounts in various organs of the aged body.
lordosis an exaggerated lumbar curvature seen in the toddler period.
lumen the opening or diameter of a vessel.
malnutrition poor dietary practice that results from the lack of essential nutrients or from the failure to use available foods.
malocclusion a malposition or imperfect contact between the upper and lower teeth.
mammography breast x-ray for diagnostic screening.
Mantoux skin test intracutaneous test for tuberculosis.
maturation the unfolding of skills or potential regardless of practice or training.
meconium the first newborn stool, usually green-black in color and odorless.
melanocytes pigmented skin cells.
menarche the onset of menses or first menstrual period.
menopause the cessation of menses, usually between 45 and 55 years of age.
milia small clusters of pearly white spots found mostly on the infant's nose, chin, and forehead. They result from nonfunctioning or clogged sebaceous glands.
molding an elongation or overriding of the cranial bones during passage through the birth canal.
mongolian spot a bluish-black, flat, irregularly pigmented area found in the lumbosacral region in darker pigmented infants.
morula the mass of cells following fertilization which resembles a mulberry.
negativistic behavior negative, rebellious behavior seen in toddlers caused by frustration or a conflict of wills.
neonate the newborn or the first four weeks of extrauterine life.
nephron the functional, working unit found in the kidneys.
normal physiological weight loss a loss of 5 to 10% of birth weight occurring in the early neonatal period with a regain in approximately 10 days.
neurons also known as nerve cells.
nurturance the provision of love, care, and attention to each family member.
nystagmus unequal eye movements (crossing) owing to immature cilliary muscles.
obesity defined as having 20 to 30% excess weight.

occult blood hidden or invisible blood.
Oedipus complex a young boy's sexual attraction for his mother and unconscious wish to replace his father.
old begins at age 75 to 90.
omnipotence sense of unlimited power or authority.
opacity a clouding of the lens of the eye.
ossification the hardening process whereby bone tissue gradually replaces the soft cartilage tissue.
osteoporosis a disorder characterized by a decreased bone mass resulting from the loss of minerals from the bones.
ova the female sex cell.
ovaries the female sex glands or gonads.
ovulation the rupture and release of the ovum.
Papanicolaou test a common screening test used to detect cancer of the cervix.
parallel play play style typical of toddlers whereby they play alongside each other but not really interacting or sharing.
penis the male sex organ.
peristalsis wavelike muscular movement found in the gastrointestinal tract.
personality the unique behavior patterns that distinguish one person from another.
physiologic jaundice a yellowish tinge to the skin of the newborn seen in the first 48–72 hours after birth.
placenta a flattened circular mass of tissue attached to the inner uterine wall. This organ has several functions: producing hormones, transporting nutrients and wastes and protecting the unborn from harmful substances.
preadolescence or **puberty** a time of rapid growth ending with reproductive maturation.
presbycusis an impairment of high frequency hearing associated with advancing age.
presbyopia the decreased ability to see objects clearly at close distance, occurs with advancing age.
primary sex characteristics the growth and maturation of the sex glands.
procreation the ability to reproduce.
progesterone a female sex hormone.
proximity nearness in location.
proximodistal a directional term that refers to growth and development that progresses from the center of the body towards the extremities.
pruritus a term used to describe itching.
pseudomenstruation a slight blood-tinged vaginal discharge that may appear shortly after birth which is thought to be related to the maternal hormones. This disappears without any treatment.
puberty the period following childhood and before adolescence in which the body prepares for the changes necessary for reproduction.
reaction time the speed at which a person responds to a stimulus.
recessive genes genes for inherited traits that can only be transmitted if they exist in pairs.
reciprocity moral feelings of concern for what is fair to others as described by Kohlberg.
regression a return to an earlier stage of development during stressful periods.
reminiscence a process of remembering and discussing key life events.
residual volume the amount of air left in the lungs after forceful exhalation.
resistance exercise weight lifting that builds muscle mass.
respectability emphasizes role modeling and valuing.
ritualistic behavior repetitive behavior, habits, or routines, that serve to lower anxiety.
saturated fats fats that come from animals such as in meats and dairy products.
school phobia an intense fear of going to school.
scoliosis abnormal lateral curvature of the spine.
scrotum a sac that holds the testes.
secondary sex characteristics all the changes that play no direct role in reproduction such as appearance of pubic, axillary, and facial hair; increase in activity of the sebaceous and apocrine glands; breast development; and a widening of the pelvic and hip bones.
senescence the symptoms or changes associated with normal aging.
senile lentigo common, flat, discolored "age spots" found on the skin.
separation anxiety anxiety brought on by stress when the young child is separated from the family by school, hospitalization, or family death.
sexuality a broad term that includes anatomy; gender roles; relationships; and a person's thoughts, feelings, and attitudes about sex.
sexually transmitted diseases (STDs) diseases that are transmitted through sexual intercourse.
sibling rivalry jealousy of siblings that causes feelings of insecurity.

socialization the process of having the individual adapt to social customs.
socializing agent the family serves as the first socializing agent for the child.
somatic pertaining to the body.
sperm the male sex cell.
stagnation the lack of generativity, characterized by having feelings of self-absorption and a general dissatisfaction with life.
stress anything that upsets psychological or physiological balance.
sun protection factor (SPF) the time a person can stay exposed to sunlight without burning.
superego the part of the mind that dictates right from wrong and is similar to the conscience.
sutures thick bands of cartilage which separate the infant's skull bones.
team play an advanced style of play typical of older children which requires the ability to follow rules and regulations. Team play may be competitive in nature.
teratogens chemical or physical substances that can adversely affect the unborn.
testes the male gonads.
testosterone the male sex hormone produced by the interstitial cells of the testes.
tinnitus a ringing in the ears.
umbilical cord the connecting link between the fetus and the placenta.
unsaturated fats fats that are usually liquid at room temperature and are derived from plant sources.
vernix caseosa a white, cheese-like, protective, covering found on the neonate's skin.
very old the period from age 90 on.
vital capacity the ability of the lungs to move air in and out.
weaning the gradual substitution of the cup for the breast or bottle.
wear-and-tear theory theory suggesting that after repeated injury, cells wear out and cease to function hastening the aging process.
wellness a relative state of health.
xerostomia a reduction in saliva and a resultant drying of the mouth.
young old the period from age 65 to 74.
zygote a fertilized ovum.

Index

Note: Page numbers followed by f indicate figures; page numbers followed by t indicate tables; and page numbers followed by b indicate boxed material.